To Edwin

Jean Tole

IS SCIENCE A DIRTY WORD?

How the Atom Bomb was made
Nuclear Power
Radio and Television
Lasers and Superconductivity
The Ozone Hole and the Environment

Ian Toler

IS SCIENCE A DIRTY WORD?

ISBN 1 85863 052 5

First Published 1994 by
MINERVA PRESS
2, Old Brompton Road,
London SW7 3DQ.

Printed in Great Britain by
Martins Printers, Berwick upon Tweed

IS SCIENCE A DIRTY WORD?

How the Atom Bomb was made
Nuclear Power
Radio and Television
Lasers and Superconductivity
The Ozone Hole and the Environment

ACKNOWLEDGEMENTS

I am indebted to Terence Miller, one time head of the North London Polytechnic, for covering my first draft with copious red ink which set me on the right lines and to my tutor, the late Dr C. H. Collie, for supporting me in the project. Also to Alan Wood who explained the intricacies of preparing work for the publisher, and to "Thompson and Thomson" for the ripple pictures on the cover. To Charles Strafford who, as a layman, made many pertinent comments and to many others for their help but especially to my daughter Celia without whose enthusiasm and support as an author herself I should never have completed the task.

CONTENTS

INTRODUCTION

"Could a Chernobyl happen here?"
"Will the Ozone Hole cause us to die of Cancer?"

When you read these headlines, do you know how atomic fission takes place or do you even know what ozone is? It is because so many of us are so uncertain about the modern science which affects our lives that I have written this book. From the sub-title it would appear that I am trying to cover far too many subjects in one short book. I am not trying to write a scientific treatise on them, and any scientific reader may find them trivial, but I would like to give the reader sufficient information for him or her to understand these matters.

In the modern world life is so bound up with science and scientific achievements, that those who have never studied any scientific subject are at a loss when they have to make a considered judgement on many of our present problems.

I hope that the reader who has begged, borrowed, stolen, or even bought (!) this book will want to learn something of the science which has brought about the atomic bomb, nuclear power and the proliferation of discoveries which affect our everyday lives and our environment. People often say "Look at those power stations (generally coal or oil fired) putting out all that pollution", pointing to the cooling towers spreading out not very harmful water vapour (steam) into the atmosphere which they think is causing acid rain, when it is the colourless gases from the chimney stack which are to blame.

They are also worried about the ozone hole and global warming but still use their motor cars because at present there is no viable alternative but not realising how much pollution they cause.

Sadly, it appears that the majority are totally ignorant of these matters. They may not know an "erg" from an "ohm" and may not want to but if they are to make reasoned decisions on subjects which will affect them and future generations, they must first understand the fundamental facts.

What has been discovered cannot be undiscovered. Canute could not turn back the waves but humankind can try to direct them into a more benign form.

There are many controversial questions, not least that of nuclear power generation which I shall consider in some detail but I have to admit that, in my opinion, it is the only practical source of power at present available in the quantity required which does not pollute the environment. This potential damage to our environment is another subject about which there is much discussion and which I shall deal with in one of the later chapters.

It is difficult for a scientist, albeit an old one, to write for the non-scientific, but intelligent reader, but it is a challenge which I have undertaken and I hope that the reader will be enlightened and I hope interested in it.

Science as the dictionary describes it, is knowledge - systematic knowledge and scientists produce accurate measurements and the understanding of phenomenon, as opposed to opinions, and they test the soundness of their conclusions. However they sometimes produce results which, although accurate and producing an increase in the knowledge of the universe, are unacceptable to us as human beings living on this planet. The atomic bomb is perhaps the best example of this in spite of the many benign uses that the study of nuclear physics has produced, such as in medicine. Gunpowder, reputed to have been invented by a German monk (scientist?) in 1300 may have killed more people than hunted animals, and we can blame the scientists for producing the chemicals and the technology causing the ozone hole and global warming and the poison gases which killed and maimed so many people in the wars.

In the early centuries if the results of the scientists were not acceptable to their rulers or religious masters they were liquidated. Copernicus kept his views on the heavenly bodies private for fear of being branded a heretic, Archimedes was killed by a Roman soldier for carrying out his experiments. Richard Dawkins, if he had lived in the sixteenth century, and written and spoken as he has on religion and science would probably have been burnt at the stake! Happily we do not do that now as it would not be very constructive, but still science if not properly understood is anathema to some people.

The science I am going to talk about is mainly physics which deals with matter and energy in the natural world. The word physics is derived from the Greek word "phasika" which can be translated as "natural things" and so physics is to do with the natural world and at some universities comes under the heading of Natural Science.

I cannot emulate such writers as Stephen Hawkins or Roger Penrose and the many others who have joined the debate, but in the last chapter I have tried to draw together some views on the origin of life, consciousness and the philosophy of reconciling science and religion.

CHAPTER I

THE FUNDAMENTALS

From the beginnings of civilisation, man has wondered what the material he handles is made of. If he grinds a piece of marble he will get a fine powder but no matter how fine he grinds it there is always something left. He may also ask, what are *we* made of? Bone, flesh and blood. But what are these made of? Up to a hundred years ago everything was considered to be made up of innumerable small particles which were called *atoms*. Sir Isaac Newton envisaged them like billiard balls in perpetual motion, striking each other and, if they were a gas, striking the walls of the vessel containing them producing a *force* which we call *pressure*.

The first difficulty we encounter in considering what we and the world are composed of, is to comprehend the smallness of the atoms from which we are made. If you take the atom in a diamond, for example, and assume it to be a sphere, its diameter is approximately one tenth of a millionth of a millimetre. A millimetre is about twice the width of a printed "l" in this sentence. Just imagine it to be split into ten million parts!

How can we visualise this? The thickness of a human hair is about one tenth of millimetre and the atoms from which it is made are a million times smaller.

But the atom is by no means the smallest particle to be considered. At its centre is a *nucleus* surrounded by even smaller particles called *electrons*. The nucleus is ten thousand time smaller than the atom and the electron is ten times smaller still. If we consider the space taken up by the nucleus and the electron in the atom we find it is so small that the inescapable conclusion is that practically all of the atom and hence all matter, including ourselves, is nothing but space. I think it was put very clearly by Sir Bertrand Russell who pointed out that when you are standing on the earth there are forces from the electrons and nuclei acting from your feet which just balance the force of gravity and this is what prevents you falling through the earth which, *solid as it looks, is almost all empty space*.

You may well ask at this point, how can we possibly know this to be true when we cannot see or feel a single atom let alone a nucleus or an electron? Well, only by the effect atoms can have on something we can observe or measure. We cannot see the wind but we can hear it and see the devastation it can cause. We cannot see or hear or feel the radio waves which are all about us but we can still hear the music from our "Walkman". To prove that atoms, nuclei and electrons do exist and to measure them we have to have recourse to theories and mathematical functions. I am not going to use mathematics as it is a mystery which only those who have studied pure mathematics can be expected to understand. However there are certain phenomena which we can observe and measure experimentally which prove these theories to be correct and from which we can deduce the properties of these fundamental particles.

Fig. 1 C.T.R. Wilson Cloud Chamber

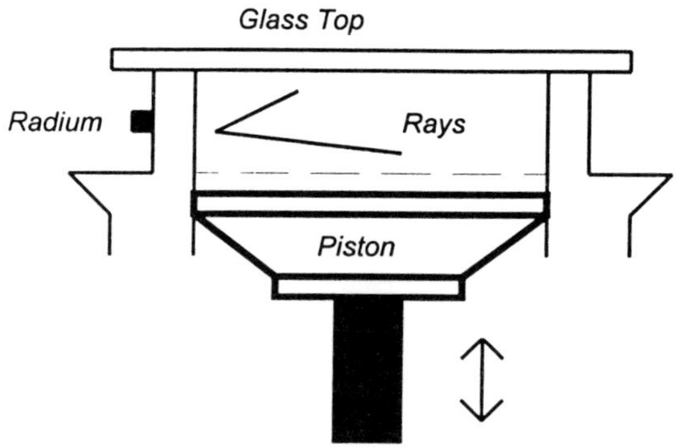

The piston is suddenly dropped from dotted line to position shown in diagram. Damp air in chamber chilled and fog forms on tracks of alpha particles ejected from piece of radium. Viewed through glass top.

One of the earliest "ways" to "see" atoms became possible in 1899 in an apparatus known as the "Cloud Chamber" invented by C. R. T. Wilson (Fig. 1). When there is enough moisture in the air and the weather gets colder, fog begins to form as we well know, making motoring extremely hazardous. In the days when most people had coal fires to heat their homes the atmosphere was polluted with smoke and particles of soot; fog formed more readily - we called it a "peasouper". In Wilson's cloud chamber the air is saturated with moisture and the atomic particle from a radioactive source such as radium is shot into it just as the pressure in the chamber is reduced by means of a mechanically operated piston. This causes the air to become what is known as supersaturated. As a result fog will tend to form if there is something for it to form on like the smoke or soot from the home fires. The atomic particle during its passage through the chamber will strike the air particles and will give them an electric charge. This helps the fog to form on them and the track of the atomic particle can be seen and photographed. The important point is that these tracks can show where particles have collided and shot off in different directions. It is from the measurement of the angles and the length of these tracks that the weight and speed of the particles which produced them can be calculated.

There are going to be a number of things we are going to talk about before we can start on the atom itself and I will explain them in as simple a way as possible.

Waves

The simplest idea of waves is probably those seen on the surface of water in a pond (Fig. 2, centre pages). We know what they look like - they have a crest and a trough - and if you fix your eye on a certain point they appear to pass it at a certain speed (Fig. 3). The distance between two adjacent crests or troughs is known as the *wavelength.* The *frequency* of the wave is the time taken for one crest or trough to become the next one and is measured in cycles (the number of times a wave goes through a crest or a trough) every second - cycles per second. If you multiply the frequency by the wavelength you will obtain the speed with which the wave is proceeding. When we are considering radio waves or even light waves, it is more difficult to visualise as they pass through a vacuum or space and there is no material such as the air or water through which they can be seen to travel. We only see their effects with the result that we can, for example, measure their speed, frequency and wavelength and so, in a way, we have to take it on trust that they do actually exist!

Fig. 3 Wave Motion

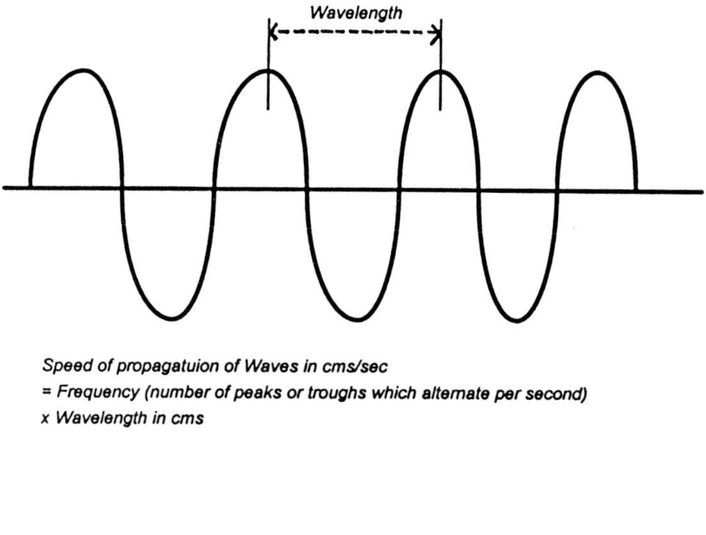

Speed of propagatuion of Waves in cms/sec
= Frequency (number of peaks or troughs which alternate per second)
x Wavelength in cms

Here are some examples of frequency and wavelength which may be of interest:

The sound of the musical note "A" vibrating in air is:

440 cycles per second

(Note - it is the air which is transmitting the sound. In a vacuum no sound would be heard.)

Take the case of radio waves:

Long Waves	-	1500 metres	-	200,000 c.p.s.
Medium Waves	-	375 metres	-	800,000 c.p.s.
Very High Frequency (VHF)	-	3 metres	-	100 million c.p.s.

Instead of cycles per second we now use "Hertz" named after Professor Heinrich Hertz, the discoverer of radio waves.

1 Hertz (Hz)	=	one cycle per second
1 Megahertz (Mhz)	=	one million hertz
1 Gigahertz (Ghz)	=	one thousand million hertz

The approximate wavelengths of various types of radiation is given in Table 1 which gives you an idea of the enormous difference in wavelengths throughout the whole spectrum. (Instead of metres one normally uses nanometres - nm - which are one millionth part of a millimetre - to avoid using too many noughts!)

All the above radiation apart from sound waves, is in the form of electromagnetic waves. You may well ask - why are they called electromagnetic waves? Not surprisingly because they consist of waves of energy associated with electric and magnetic fields and resulting from the acceleration of an electric charge. Further discussion on these waves which pervade all space will be left to a later chapter.

Table 1

Wavelength and Frequencies of Electromagnetic Waves

(Note the very small range of visible light)

Approximate Wavelengths in Metres		Approximate Frequencies in Kilo (thousand) cycles/sec
10^{-17}		10^{22}
10^{-16}	⇑	10^{21}
10^{-15}	Cosmic Rays	10^{20}
10^{-14}		10^{19}
10^{-13}		10^{18}
10^{-12}	Gamma Rays	10^{17}
10^{-11}		10^{16}
10^{-10}		10^{15}
10^{-9} (1 n/m)	X-rays	10^{14}
10^{-8}	Ultra-violet Rays	10^{13}
10^{-7}	Visible Light	10^{12}
10^{-6}		10^{11}
10^{-5}	Infra-Red Radiation (Heat)	10^{10}
10^{-4}		

		10^9
10^{-3}		
	Microwave	10^8
10^{-2}		10^7
10^{-1}	Radio	
		10^6
1 metre	VHF - Very High Frequency	
		10^5
10	HF(FM) - High Frequency	
	(8.8 to 10.8 x 10^4 k/c)	10^4
10^2	MF(MW) - Medium Wave	
	(530 to 1600 k/c)	10^3
10^3	LF - Long Wave	
	(160 to 260 k/c)	10^2
10^4		
	VLF - Very Long Wave	10
10^5		
		1

Sound "A" 440 cycles/sec. Ultrasound above 20,000 cycles/sec
1 Metre = 1,000,000,000 Nanometres
10^{-17} = 0.000,000,000,000,000,01

Forces

Fig. 4 *Magnetic Force*

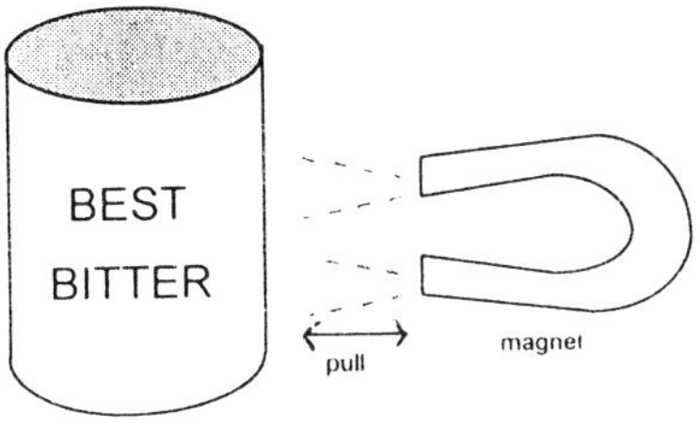

TINPLATE (STEEL) BEER CAN
NOT ALUMINIUM!

(you can tell the difference - no PULL with aluminium!)

So far we have been considering the size of the atom and its components. Before we go into its structure we must consider what we mean by *force*. We cannot see force but we can feel its effects. If someone punches you on the nose, you can see and feel the effect even though you cannot see the force. If you hold a steel magnet in your hand, you can feel it being pulled towards another piece of steel (Fig. 4). The force pulling it is called *magnetic force*. You will notice that it can be felt even at some distance from the steel although it is much stronger when it gets really close. It doesn't matter too much what is in between the magnet and the steel. All of the forces we shall discuss "act at a distance" and are only affected by the material in between to a certain degree. Our blow on the nose would be reduced if we were wearing a baseball helmet!

When you brush your hair on a frosty day and if the air is very dry, you will notice that your hair is trying to stand up straight and is being attracted by the brush. A more modern everyday example of this is when you walk briskly across a carpet which nowadays is frequently made of nylon. If you are wearing rubber soled shoes you will generate an electrical charge on yourself which will produce a small spark if you touch another object. My granddaughter was always pleased to demonstrate this by coming up behind me unsuspectingly and producing a spark from my neck causing me to jump! What is happening is that an *electrostatic charge* is being produced on the hairs by the friction caused by the brush. This charge can be of two kinds, described as being of positive or negative polarity. Charges of similar polarity repel each other and those that are different will attract, as we shall see when we describe electricity. The process of brushing causes the hairs all to have one polarity so they repel each other and try to stand upright. The brush has the opposite polarity and not only will it attract the hairs to it, but if the charges are strong enough they will cross the air gap between the brush and the hair, causing a small spark which you will hear as a "crackling". If it is quite dark you may even see the sparks, so what you have here is an example of the *electrostatic force*.

Now if you were to wind a coil of copper wire round a cardboard tube and pass an electric current through the wire and if you then held a steel rod at the end of the tube, you would find that the steel rod would be pulled into the tube. This is because the electric current produces a magnetic field similar to the one produced by the steel magnet which we have just described. This force caused by a flowing electric current is called the *electromagnetic force*.

Finally there is the force of *gravity*. We all know, as Sir Isaac Newton is reputed to have found to his cost, that an apple falling from the tree to the ground struck his head with some force! When you try, if you are a farmer or a greengrocer, to lift a sack of potatoes from the ground, you find there is a very strong force trying to stop you doing so. If you are not careful you will end up with a "bad back"! This force is the *gravitational force*. Although, perhaps, not fully understood, it is the universal force of attraction between all matter. Its strength is proportional to the masses (the quantity of matter in the bodies) and inversely proportional to the square of the distance between the masses, i.e. the distance multiplied by itself. This is the "inverse square law" which we shall come across with many but not all the forces we are concerned with.

There is an important distinction between mass and *weight*. The weight of the sack of potatoes is dependent on the force of gravity between it and the Earth, so its weight will diminish as the distance from the Earth increases. As a result, if you took it up in a spacecraft far enough from the Earth, it might weigh nothing and you could toss it over your shoulder with your little finger. The weight of a body could be described as the force of attraction of the Earth on it, although on the Moon it would be quite different. Its mass remains the same and does not vary, but we shall see later that when we are thinking of very small bodies moving at very high speeds, approaching that of light, the mass no longer remains the same with very devastating results!

These then are the main forces we shall have to consider. There are other ones which we shall meet when we get inside the nucleus of the atom but they need not concern us at the moment.

Electricity

Six centuries before Christ, Thales, a Greek philosopher and astronomer knew that when amber was rubbed with silk it acquired the property of attracting light objects. From the Greek word for amber - "elektron" - the word electricity has been derived.

Fig. 5 Franklin and Kite

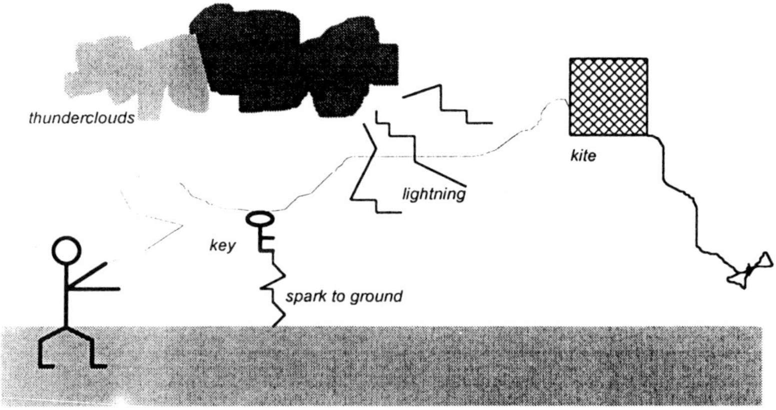

In 1752, in a field near Philadelphia, Benjamin Franklin flew a kite in a thunderstorm and obtained a spark from the end of the line he was holding (Fig. 5). Certainly not an experiment to be recommended as Franklin was probably lucky not to have been killed. Michael Faraday proposed the first theory of electricity. He recognised that there were two kinds of electricity - positive and negative - that attracted or repelled each other according to the inverse square law we have just described. In case this is not clear, so let me explain. Suppose you have two charges one positive and one negative two centimetres apart and suppose you then halved the distance between them to one centimetre. You might think the attraction between them would be twice what it was before but in fact it will be four times. The difficulty that Faraday and others had to explain was how these forces acted at a distance to each other when there was *nothing* in between. The idea of a universal "aether" was therefore introduced but over the years it has never been possible to prove that it exists so we are left with the fact that electric charges do act on each other at a distance with nothing in between. We shall return to this later.

There were many theories of electricity produced in the latter half of the nineteenth century and the word "electron" for the elementary electrical entity was first suggested in 1891. The amount of the electrical charge on this particle called the electron has been calculated and later confirmed by experiment. When an electric current flows in a wire such as the electric cables in your house, the unit of current which we can consider as a flow of electrons through the wire, is called the "ampere". Whilst the ampere is the measure of the flow of electrons in the wire the fundamental unit of electricity is the "coulomb". This is *not* the quantity of electricity shown on your bill from the electricity board which takes into account both the current and the voltage (or one might say the "pressure" of the electricity) and the time for which it has been flowing. It is shown on your meters as Kilowatt-hours. A single bar electric fire generally takes 4 amperes at 250 volts which is 1000 watts or 1 kilowatt (1Kw). So it uses 1 Kw/hour or 1 unit every hour that it is on so you can work out the cost!

The charge or quantity of electricity on the electron is minute - 1.6 of a coulomb divided by 10 with 18 noughts! In order for a current of 1 ampere to pass in 1 second, it would require a number of electrons equal to 6.25 multiplied by 10 with 17 noughts after it (remember one million is 10 with only 5 noughts after it!) When you consider that a 100 watt light bulb passes a current of about half an ampere you can get some idea as to how small the charge on an electron is - but remember there are an awful lot of them!

I'm sure you must now be thinking how on earth can we measure such remarkably small quantities. In 1897 J. J. Thomson, working with an apparatus called a cathode ray tube - the forerunner of the TV tube in our "telly" - discovered a negative corpuscle which was produced irrespective of the gas in the tube and whose charge he was able to measure - this we now know was the electron. I would like to explain the cathode ray tube but before doing so I think I should explain another fundamental unit of electricity - the "volt". When electricity flows in a wire there is a certain resistance which tries to stop it passing. I will explain later how this comes about but for the moment 1 volt is the pressure or voltage required to pass 1 ampere through a wire which has a resistance which is defined as 1 "ohm".

Fig. 6 Cathode Ray Tube

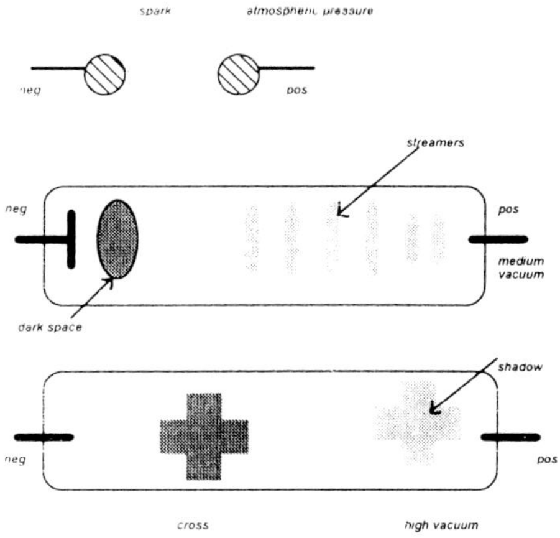

Now to describe the cathode ray tube (Fig. 6). If the electrical potential or voltage between two pieces of metal which we will call electrodes (one the cathode or negative electrode and one the anode or positive one) and which are in air, is increased, then, depending on the distance apart of the electrodes and the pressure of the air, a discharge of electricity in the form of a spark will occur. This is what takes place in the sparking plug of your car (if it has a petrol engine rather than a diesel) the voltage in this case being of the order of several hundred volts. If you now seal your electrodes some distance apart in a glass tube from which the air is gradually removed, the discharge is no longer a spark but assumes a silent streamer like form and becomes coloured depending on the nature of the gas in the tube. In the case of air - pink. If it is neon (one of the rare gases present in the atmosphere) - reddish (hence "neon" signs) - sodium vapour - yellow (the yellow street lights are basically cathode ray tubes to which a small amount of a sodium compound has been added). This type of discharge occurs when the air in the tube is about one half atmospheric pressure. As the pressure is reduced further, a dark space appears round the negative electrode, (the cathode) and at a pressure of about .01 mm (normal atmospheric pressure is about 760 mm) a dark space occupies almost the whole of the tube. Although at this pressure nothing is visible in the tube, the tube itself glows with a bright fluorescence, the colour depending on the type of glass or on a coating which may have been put on the inside of it. For example; soda glass - bright green, lead glass - blue, a coating of corundum (aluminium oxide) - bright crimson, zinc sulphide - bright blue.

This is the basis of our fluorescent light tubes. These tubes contain a cathode which is heated electrically and gives off electrons as in the cathode ray tube. A small amount of mercury is placed in the tube and when the electrons strike the mercury atoms ultraviolet radiation is produced. This radiation is invisible but the inside of the tube is coated with a compound which fluoresces brightly when struck by UV radiation and produces the light we see. UV radiation is considered to be slightly harmful as we shall see in Chapter VII but the glass of the tube and the coating prevents it from escaping. Mercury is also poisonous as is the coating inside the tube. This is why care should be taken in disposing of old tubes especially if they are broken.

If a metallic screen, say in the form of a cross, is built into the tube in front of the cathode, its shadow will be thrown on the end of the tube, indicating that whatever is being emitted is travelling in straight lines. In the same way by placing a collecting box or second electrode in the path of the rays and connecting this to an instrument for measuring an electric charge it was shown that whatever the rays consisted of, they carried a negative charge. It was therefore concluded that whatever was being emitted from the cathode and propelled towards the anode were electrons and these rays were known as "cathode rays".

An experiment was next tried of placing a platinum plate inside the tube so that these rays would strike it. A new form of radiation appeared which was found to be capable of passing through many solid substances and would then affect the emulsion on a photographic plate. These new rays were first discovered by Rontgen in 1895 and were called "X rays" (Fig. 7). They are similar to light and UV radiation but of very much shorter wavelength, and also to "gamma rays" which we shall discuss under radioactivity. As we all know they pass through flesh but are stopped by bone so that when you go to the dentist he can take a photograph of the roots of your teeth and tell you if you have to have one out. The penetrating power of X rays depends to a great extent on the degree of vacuum in the tube and to the voltage between the electrodes - the higher the vacuum (i.e. the less the pressure) the more penetrating or "harder" the X rays.

So that is the cathode ray tube which I have always thought was one of the most fascinating pieces of equipment used in the early days of probing into what we are made of.

Fig. 7 'X' Ray Tube

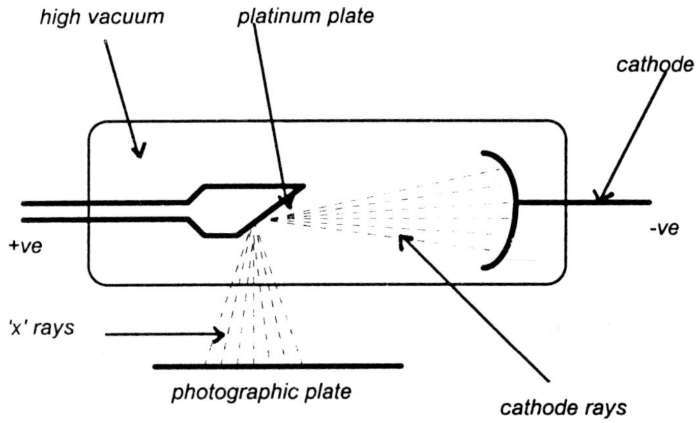

Ions and Ionisation

When you come across these words you may wonder what they mean, so, as they will occur frequently we had better try to find out more about them.

Let us assume we have two electrodes - an anode (positive) and a cathode (negative) similar to those we have just described in the cathode ray tube, but placed in a liquid such as a solution of common salt in water. If we now connect a source of electricity to them, say from a battery, an electric current will flow from the anode to the cathode through the solution. But the atoms in the solution, as we shall see later consist of a positively charged nucleus together with a number of negatively charged electrons just balancing this positive charge, and are therefore neutral. So how can a current flow when there are no free electrons to carry it? (Fig. 8.)

Fig. 8 Ions in Solution

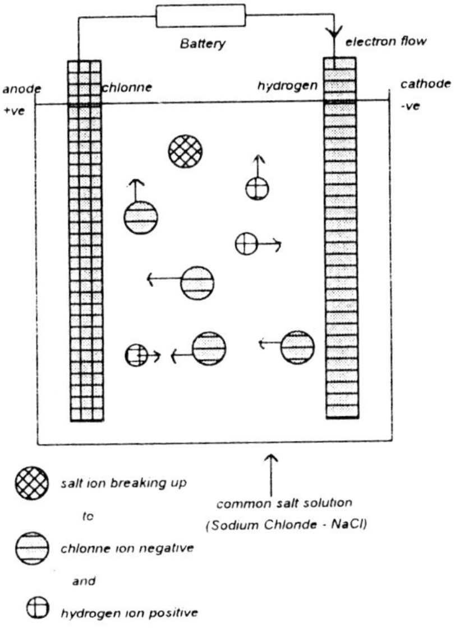

The answer is that in a solution, and especially in a dilute solution, some atoms lose some of their electrons and become what are called "*ions*". Some of these ions are positively charged as they have lost some of their electrons which had originally balanced the charge on the nucleus, and some will gain electrons and become ions which are negatively charged. *Overall* the solution is still electrically neutral. However when an electrical potential (voltage) is applied to the electrodes the positive ions are attracted to the negative electrode and the negative ones to the positive and a current will flow.

Common salt is composed of sodium and chlorine atoms - sodium is a very reactive metal and chlorine a greenish yellow gas used in the First World War as a chemical weapon. When these atoms combine to form the benign common salt it has none of the unpleasant properties of its constituent atoms and is called a *molecule* of sodium chloride. We might here define a molecule. This is a group of atoms bonded together to form a particular substance and there may be many hundreds or thousands of them in some molecules, especially those which make up our bodies.

To return to our solution. The sodium ion which is positive moves towards the negative electrode (cathode) whilst the chlorine ion which is negative moves towards the positive electrode (anode). What happens then is that the chlorine is released as a gas at the anode. The sodium is very reactive and when it reaches the cathode it immediately reacts with the hydrogen in the water to form sodium hydroxide or caustic soda. You may be interested to know that this process is used on an industrial scale to produce large quantities of caustic soda, hydrogen and chlorine, which are the basis of many industrial and household products. The process is called "electrolysis" and illustrates the formation of ions in solutions.

In the case of gases the situation is slightly different. Normally they do not conduct electricity unless the pressure is reduced and the potential increased as we have seen in the cathode ray tube. If the temperature is raised or the gas is subjected to radiation such as X rays or UV light some of the atoms become "ionised" - some losing electrons and some gaining, with the result that the gas will conduct in much the same way as for a liquid.

It might be appropriate to digress for a moment at this point on the conduction of electricity in solids. Consider an electric current passing through a copper wire. If the conduction were to be caused, as in liquids and gases, by the movement of ions, then we would expect an increase in the amount of copper at one end of the wire and a diminution at the other. This just does not happen (or all the wiring in your house would collapse). The explanation is that in solids the conduction is due to the movement of *free* electrons which are present in conductors such as copper but not in non-conductors such as glass (Fig. 9). When a voltage is applied to the wire, the electrons will flow towards the positive end and you can imagine them being pushed round to the negative end by the battery or generator (although the direction of the flowing current is from negative to positive - the "conventional" direction is still spoken of as positive to negative).

Fig. 9 Conduction of Electricity in Copper

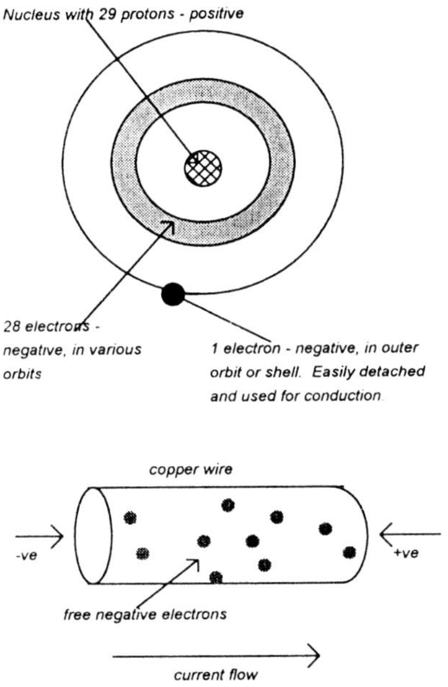

The question of orbits shown in Fig. 9 will become clear in Chapter II.

Periodic Table

To try to weigh the atoms of each element would be very difficult in view of their small size. However what we do is to compare their weights with a standard using hydrogen as 1 and oxygen as 16. If we now assemble these weights, which we call the *atomic weights* of the elements, in ascending order from hydrogen as one, we obtain a list which I have shown in the Appendix which includes all the elements with their number which is known as the *atomic number*. (The significance of this number will appear later.)

When these numbers are studied, it was pointed out in 1896 by a Russian chemist Mendeleef that there was a periodicity in them. Elements with similar properties recurred at equal intervals. If put in a table, for example - lithium, sodium and potassium had numbers 3 - 11 - 19. The difference between each number being 8. Another group - boron, aluminium, gallium had numbers 5 - 13 - 31. Again with a difference of 8. As you go further up the table this simple correlation is no longer correct but there is still a periodicity. I have shown a portion of this table which is known as the *periodic table*, in Table 2. It was also noticed that there were gaps in the numbers in the table which belonged to no element known at that time. Subsequently these elements have been discovered and not only that, but they have the properties that had been predicted. Originally the highest number in the table was 92 with hydrogen 1 and uranium 92 and several gaps. These gaps have all been filled and the table extended to 103 with the element lawrencium.

Table 2

The Periodic Arrangement of the Elements up to No. 25

The Atomic Number is given followed by its symbol and atomic weight.

Hydrogen 1 H 1.008							Helium 2 He 4.003
Lithium 3 Li 6.940	Beryllium 4 Be 9.020	Boron 5 B 10.820	Carbon 6 C 12.010	Nitrogen 7 N 14.007	Oxygen 8 O 16.000	Fluorine 9 F 19.000	Neon 10 Ne 20.183
Sodium 11 Na 22.997	Magnes-ium 12 Mg 24.320	Alumin-ium 13 Al 26.980	Silicon 14 Si 28.090	Phosph-orus 15 P 30.975	Sulphur 16 S 32.064	Chlorine 17 Cl 35.453	Argon 18 A 39.944
Potassium 19 K 39.100	Calcium 20 Ca 40.080	Scandium 21 Sc 44.960	Titanium 22 Ti 47.900	Vanad-ium 23 V 50.950	Chrom-ium 24 Cr 52.010	Mangan-ese 25 Mn 54.930	
							Krypton 36 Kr 83.800

Apart from hydrogen and krypton all the atomic numbers in the vertical columns differ by 8 and those elements have similar basic properties. The table can be extended but becomes more complicated with various sub groups.

We have already seen that the atom is made up of a nucleus with sufficient positive charges to balance the negative electrons outside it. If we look at the Appendix it is apparent that the atomic weights of the elements are not always whole numbers, i.e. 1 - 2 - 3 - 4 etc., but fractions. For example chlorine is 35.457. However if we take the *number* of unit *charges* in the nucleus we do get whole numbers as the electron has the smallest charge that can be envisaged - fractional charges just do not exist. This number of unit charges on the nucleus is the *atomic number* which is the number shown in the table in the Appendix.

Now we have to explain why the atomic weights are not whole numbers. The answer is that some elements can exist with a mixture of atoms with different atomic numbers. For example the chlorine atom consists of a mixture of chlorine atoms of atomic number 35 and 37 mixed in the approximate proportions of 3 to 1 which makes the atomic weight of the normal chlorine atom 35.457. If oxygen is taken as 16 then hydrogen is not I but 1.008 and uranium 238.03 both facts which we shall find very important when we come to atomic fission and fusion.

Radioactivity

After Rontgen had discovered X-rays, which I mentioned when I described the cathode ray tube, a French physicist, Antoine Henri Becquerel, was experimenting with uranium compounds in connection with phosphorescence. Phosphorescence and fluorescence are the properties of a substance which gives off light of one colour when subjected to light of a different colour or wavelength. You may have come across a dye called fluorescine which is sometimes used to mark the flow of water when it is leaking from a pipe or drain. A very small quantity will fluoresce with a brilliant green colour in daylight, so you can trace the leak. If you are a philatelist you will know that certain stamp papers fluoresce brightly when exposed to the light from an ultraviolet lamp and can make a considerable difference to their value! Fluorescence ceases as soon as the source of light is removed whilst phosphorescence will continue for a time after it has gone. This is the basis of some luminous watch dials, the numbers being coated with calcium sulphide which is neither radioactive nor dangerous, but will show up in the dark after having been exposed to light.

What Becquerel found, during his experiments, was that not only did the uranium compounds cause phosphorescence in other compounds, but were emitting something themselves which would pass through material which was opaque to ordinary light and would then expose a photographic plate in the same way as X-rays. It was later found that thorium compounds also had this property. Thorium compounds were part of the incandescent gas mantles which were used for street lighting and in the home at one time. I wonder how many of you remember the "Aladdin" paraffin lamps with incandescent mantles in the 1920s? I had the job at home of filling and cleaning them!!

In 1898 Pierre and Marie Curie were investigating this property of uranium and succeeded in separating from uranium ore - pitchblende - a new element which they called "Radium", and another one they called "polonium" (NOT plutonium) which Mme Curie called after her native Poland. It was a very laborious process. They had to handle seven tons of pitchblende which came from the mines in Czechoslovakia, to obtain only two grams of radium. In the process, and over the years, Mme Curie suffered severely from the radiation burns to her hands and body (the effect of this radiation on the human body was not at that time fully understood).

The radiation given off by radium is in many ways similar to that given off in a cathode ray tube or an "X" ray generator in that it produces ionisation (conduction) in the gas it passes through. This radiation, known as radioactivity, is of three different kinds:

Alpha Rays

These have a positive charge, are of atomic dimensions (helium nuclei). Their penetrating power is low being stopped by a few centimetres of air.

Beta Rays

These are electrons with a velocity approaching the speed of light - over six hundred and seventy million miles per hour! They are more penetrating than alphas, losing half their intensity in one metre of air.

Gamma Rays

These are similar to X-rays but of shorter wavelength and more penetrating power, passing through one centimetre of lead before losing half their intensity.

As a result of further research it was realised that this radioactivity came from the radium *atom itself* and no physical chemical change had any effect on it. Sir Ernest Rutherford who we shall meet again, showed that the alpha particles consisted of helium atoms with two positive charges which were not cancelled by the negative electrons.

A further point before we leave radioactivity is to ask how long this property of giving off radiation lasts? It is measured by taking the time the intensity of the radiation takes to fall by 50% - known as the "*half life*" of the element. This may range from a fraction of a second in the case of radium C - to 1620 years for the radium as isolated by Mme Curie and millions of years in the case of natural uranium (or there would be no uranium left on Earth and perhaps no atomic bomb!)

Isotopes

You have probably read about isotopes in the media, especially in connection with radioactive isotopes in the environment after a nuclear explosion. I will try to explain without encroaching too much on the next chapter.

The charge on the nucleus determines the atomic number and we take the hydrogen nucleus (At. No. 1) to be the fundamental building block of all the elements. It has a nucleus with one positive charge and one negative electron outside it. The helium atom (At. No. 2), one step up the table, has a nucleus with two positive charges and two electrons outside it but its weight is *four* times that of the hydrogen atom. How can this be? If the helium nucleus is made up of four hydrogen nuclei the weight would be correct but the charge too high, i.e. it would have four positive charges and only two negative electrons so it would have a nett positive charge of two. It was then suggested (before the discovery of the neutron) that the four hydrogen nuclei might have two electrons *inside* the nucleus, which with the two orbiting electrons would make the helium atom neutral.

The helium atoms which Rutherford found in the radiation from radium (alpha particles) were helium atoms with two positive charges, i.e. they were helium atoms stripped of their two exterior electrons. This supported the suggestion.

So we now see that when a radioactive element, or one that has been subjected to radiation, emits an alpha particle, its atomic number will be reduced by two (the helium removing two charges from the original element's nucleus) and it therefore moves two places down the periodic table and becomes another element. Thus radium (Atomic No. 88) becomes radon (At. No. 86) which in turn becomes polonium (At. No. 84) and finally lead (At. No. 82).

So the alchemist's dream of the transmutation of the elements comes true, but, alas, the product is not gold but lead and at that time nothing that Man could do would influence it. It was a natural phenomenon.

If a beta particle, which you remember is an electron, is emitted, the nucleus is left with an extra positive charge which means it will move up the table by one place. It is possible however for an element to lose one alpha particle and two beta particles at the same time. It will then return to the same place in the table with the same atomic number but with a different atomic weight and is called an isotope of the original element.

As the atomic number generally determines the chemical properties of the element, isotopes are very difficult to separate chemically as we shall see later.

All this must seem very complicated but the question of the isotopes of radioactive material is an essential one if we are to understand how the atomic bomb was developed.

CHAPTER II

THE ATOM AND THE NEUTRON

The Atom

The picture of the atom in the early part of the twentieth century was of a central nucleus, consisting of positive charges, surrounded by electrons which had opposite charges. The number of electrons was equal to the number of positive charges on the nucleus with the result that the whole atom was electrically neutral. This picture however produced a number of problems.

The first which had to be explained, was how these electrons were to be distributed. If they were stationary the electrostatic forces of attraction - positive to negative - would simply pull them into the nucleus, they would disappear and there would be no free electrons, the atom would be annihilated, and we would not exist!

It was then thought that these electrons might be rotating round the nucleus in orbits rather like the planets in our Solar System. But this produced an insuperable difficulty. From the classical point of view (which means the physics propounded by Sir Isaac Newton in the nineteenth century) such an atom would be inherently unstable. The orbiting electron in its movement round the nucleus would have the same effect as an electric current which you remember is after all the movement of electrons in a wire. This would result in energy being radiated in the form of electromagnetic radiation as we have seen in chapter I. As a result the electron would lose some of its energy, slow down, eventually crash into the nucleus and disappear with the same result. But atoms are inherently stable and, apart from the radioactive elements, do not radiate energy. A further difficulty was that if you had several electrons in the same orbit, their natural repulsion would cause them to become unstable, fly apart, and the atom would again disintegrate.

These then were the problems, until there appeared on the scene a Danish physicist - Niels Bohr. He was, apart from Einstein, probably the most famous scientist to investigate the atom and to follow the development of the bomb, about which he had very great moral reservations. Bohr made the assumption that there were forces in nature completely different from the classical mechanical ones understood up to that time. In 1913 he made the highly unorthodox proposal, contrary to all the laws of classical physics, that electrons rotating in certain fixed orbits *did not radiate energy*. Only when an electron moved (jumped) from one orbit to another closer to the nucleus was energy emitted. This revolutionary theory was supported by some discoveries made previously.

Fig. 10 *Oxygen Atom*

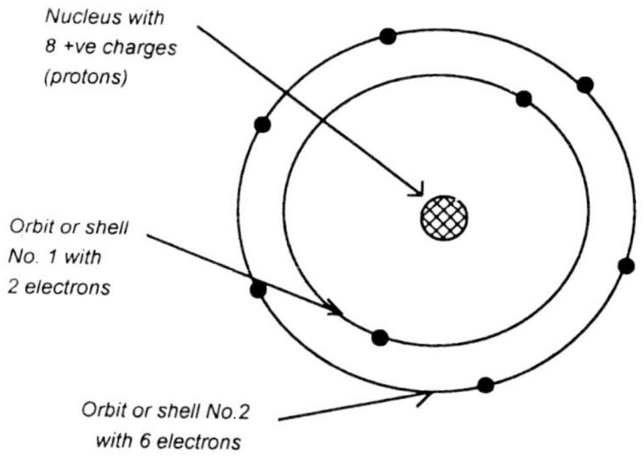

Classical theory - electrons rotating round nucleus - held in orbit by centrifugal force. They would radiate energy, slcw down and crash into the nucleus.

Bohr's theory - electrons can only jump from one orbit to another. If to a lower one, energy will be radiated. Electron just cannot exist at any point in between orbits.

In 1900 Max Planck, a Danish theoretical physicist, had proposed that a vibrating particle such as an electron could radiate energy *only* at certain frequencies. The energy arising at these permitted frequencies could be determined using a universal constant known, not surprisingly, as Planck's constant, generally denoted by "h".

The value of "h" is another of these minute numbers - being 6.63 divided by 10 with 26 noughts after it! It is measured in ergs per second. Now you are going to ask me what is an "erg"? It is a unit of energy or work done - not of course your homework if you are at school or some other chore! Although these do employ energy, either mental or physical.

When in chapter I, I was discussing forces and in particular gravitation, I didn't mention acceleration. It is important to differentiate between speed and acceleration. If your motor car is travelling at 30 miles per hour, and supposing no friction or wind resistance or anything trying to stop it, then it would require no extra energy to keep it going. However when you want to accelerate to 40 m.p.h. you must increase the energy given out by the engine by pressing the accelerator, and this amount of extra work from the engine makes the mass of the car accelerate and go faster. It is this work or energy which can be measured in ergs but the unit is *very small*. It concerns a mass of only one gram (a pound is 453 grams) accelerated through one centimetre in one second. So you see you would need a lot of ergs to move your car!

However to return to Max Planck. He stated that energy could only exist in little packets to be known as "quanta" and these could be calculated from the frequency of the radiation multiplied by "h" and a whole number - 1, 2, 3, 4 etc. This fundamental theory known as the "quantum" theory ("quantum" from the Latin "quantus" - how much?) has been proved conclusively to be true from experiments which I shall not describe although we may come across some of them later.

We might stop and think for a moment what the quantum theory means to us. It is something which is incomprehensible in the picture of the world which we see and live in. For example: a motor car accelerating, as I have suggested above, from 30 to 40 m.p.h. under quantum conditions could *never have any speed between 30 and 40 m.p.h.*, - speeds between would just not exist. You would move from 30 to 40 m.p.h. instantly - what a car! Of course this is not possible in the case of a car but at atomic dimensions mathematical calculations and experimental observations show that it can happen.

That is why it is so difficult to contemplate in our world where everything flows continuously in our life and in our mind and we are now told it is discontinuous and that some measurements of matter just cannot exist. It must leave us wondering what *is* real.

That is why I think the quantum theory and what is known as "quantum mechanics" which is derived from it, is one of the most shattering ideas that science had come up with so far.

The next support for Bohr's theory was what is known as the photoelectric effect. When light shines on certain elements (silicon for example) electrons are emitted. This is the effect by which solar powered watches, calculators etc. work and which produces the power in space craft from their solar panels. It had been found that the energy of the electrons emitted by this effect did not depend on the brightness (intensity) of the light, as might have been expected on the basis that the more you put in the more you get out, but on its colour or frequency. So ultraviolet light would produce more electrons and energy than visible light. Following on from Planck's theory, Albert Einstein had suggested in 1905 that light, which had up to that time always been considered as travelling in waves, actually travelled in small packets of waves which he called "energy quanta" or "photons". He also calculated that the photon would have an energy proportional to its frequency multiplied by Planck's constant "h". This suggestion, that light was no longer travelling in waves but as particles, was completely contrary to everything which had been thought previously, but it supported Bohr's theory that the energy emitted by the atom was discontinuous and was given off in quanta as the electron jumped from one orbit to another.

The next thing that supported the theory was the investigation of "spectra". Now to explain this we go back to the classical theory of light, namely that it is in the form of waves not particles. The colour of light depends on its wavelength - red light is of longer and violet light of shorter wavelength - white light is a mixture of all the colours and wavelengths and appears to the eye as white. If we look at the colour of a piece of green paper for example. The green pigment absorbs all the colours *except* green. Only the green light is reflected and we see green. When white light is passed through a glass prism the different wavelengths of which it is composed are bent by the prism by differing amounts depending on their wavelengths (Fig. 11). If the resulting light is projected on a screen it appears as a series of coloured bands from red to violet. This is what happens in the rainbow. Here the light from the Sun, which is behind you is reflected and as in the prism, refracted (or bent) by the raindrops which are in front of you and so form the coloured bow. If, as sometimes happens, the light is reflected twice by the same raindrop we get the second bow. This second reflection has the effect of reversing the colours in the second bow. Have a look next time you see a rainbow.

Fig. 11 *White light passes through a glass prism dispersed into a coloured spectrum*

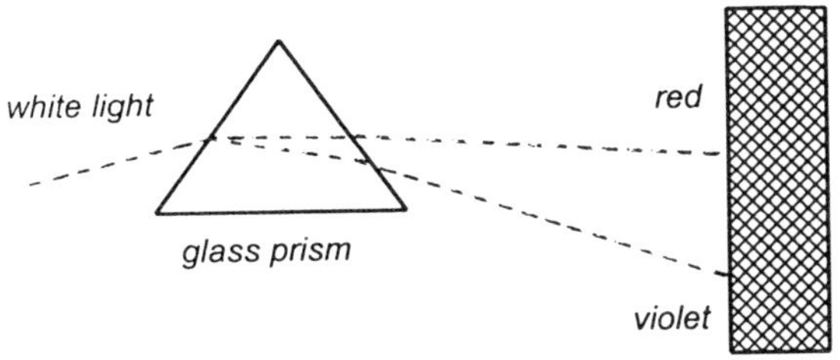

Going back to our prism. If instead of white light we use light from a chemical source such as common salt (sodium chloride) produced by heating the salt in a flame and pass this light through a prism and a series of lenses - in an instrument known as a "spectrometer" two intense yellow lines are seen. If the wavelengths of these lines are measured they are found to correspond to the frequency and therefore the wavelength which would be emitted if an electron jumped from one orbit to the next lower one in the sodium atom - an extraordinary support for Bohr's theory. More precise measurements with more accurate instruments have shown that the many lines produced in the spectra of other substances correspond with all the elements known at the time, and many which have been discovered since.

You may ask at this juncture why does an electron pick a certain orbit which is specific to the atom, and no other? The answer is a mathematical one and I have promised not to include any mathematics. However the mathematicians have considered all the possible orbits and on the basis of "probabilities" or "odds" have decided on the most probable ones, and they have said it is impossible to even conceive of an electron in between these orbits. You can think of the analogy of a roulette wheel. The ball can only enter one of the cups round the periphery. There is no possibility of it being in between. Another reason why only certain orbits are possible is when we consider the electron as a wave. The number of wavelengths must fit exactly into the orbit otherwise they would start to interfere with each other and become unstable so for each electron with a certain wavelength there can only be one orbit.

We have now come up against two fundamental concepts in atomic physics: "probabilities" and the "duality of matter". The latter indicates that particles, as we call them, can be considered as either having mass and acceleration (a particle) or as a packet of waves having frequency and wavelength (a photon - light).

Probabilities have always intrigued me - not, I assure you, on trying to decide which horse is going to win the Grand National (if it starts!) - but if all phenomena are based on a calculated mathematical probability, how can we be sure of anything. Perhaps if we had sufficiently powerful computers everything could be predicted and the future might no longer be so unpredictable! What a thought! However mathematicians such as Roger Penrose and Stephen Hawkins tell us this is "impossible". I rather hope they are right.

I would like to digress for a moment and think about the origin of life on this planet for which there are many theories apart from the theological one. We do not know what raw materials were present when the Earth was formed but it is reasonable to assume that carbon dioxide, methane and ammonia were present as these are found on other planets of our Solar System. Scientists have tried to simulate the effects of lightning and ultra violet radiation, which must have been present at that time, on these materials. They subjected them in the laboratory to ultraviolet light and high energy electrical discharges over a long period of time and found that they could produce a weak brown soup containing amongst other things, amino acids which are the building blocks of proteins - an essential part of living matter. With further inputs of energy these would combine into larger conglomerates of molecules which you may have heard of as DNA and RNA - but there was still no "life". Then it is assumed that by "chance" - a very long probability - over a very long period of time a molecule was formed which could create copies of itself. You must remember that the odds on something happening shortens with the number of times you try it. If you play the fruit machine for a very long time you will certainly win some prizes. Here we are dealing with millions of millions of years and the molecule which "replicates" itself did happen. It is now known as a "gene" and that is one explanation of why we are here.

The Neutron

As we have discussed in the previous chapter, we can consider the helium atom as being made up of a nucleus which would consist of the nuclei of four hydrogen atoms) with two orbiting electrons. But hydrogen nuclei each have one positive charge so with four electrons in orbit the helium atom would not be neutral but would have a nett negative charge. So something was wrong and had to be explained.

In 1920, Rutherford had suggested at a famous lecture that perhaps there existed in nature another particle of the same mass as the hydrogen nucleus but with *no charge*. He suggested a name for it - the "neutron". If this was so it solved the problem. The helium atom would consist of only two hydrogen nuclei plus two neutrons which would make the mass correct (four times the hydrogen atom) and the two electrons would balance the two charges on the hydrogen nuclei and the atom would be neutral (Fig. 12).

Fig. 12 Helium Atom

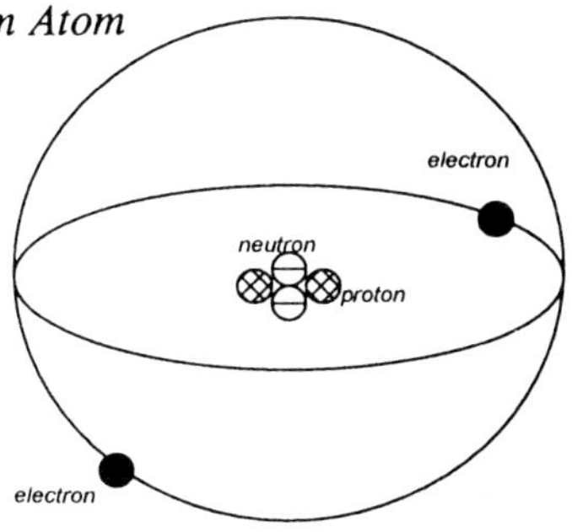

At this stage I think I should explain that the hydrogen nucleus without its orbiting electron can be considered as a hydrogen ion with one positive charge which we now call a "proton". All matter would be made up of protons and electrons and, if they could be found, neutrons. The search for the neutron was then on.

In 1919, Rutherford had been studying the effect of the radiation from a radioactive source on pure nitrogen - one of the constituents of our atmosphere. He found, to his amazement, that hydrogen atoms were being produced by the effect of alpha particles (helium nuclei) emitted by the radioactive source on the nitrogen. In writing his paper on the subject he stated:

"From the results so far, it is difficult to avoid the conclusion that the atoms arising from the collision of alpha particles with nitrogen are hydrogen... if this is the case we must assume that the nitrogen atom is disintegrated."

The Press then announced "Rutherford has split the atom". It was in fact not a split but that the first artificial transmutation of an element had been achieved. What it meant was that the nucleus of the nitrogen atom had been broken releasing one proton and one electron to form a hydrogen atom. As a result much work was carried out on attacking the nucleus of elements by bombarding them with alpha particles. However to get past the formidable electrostatic forces holding the nucleus together the bombarding particles would have to go very much faster and have much greater energy than was possible at that time.

A chemist - Francis Aston - now came on the scene. He developed an apparatus which we now call a "mass spectrograph" by which atoms of differing charge (atomic number) and mass could be identified which also meant that isotopes could now be separated. The basis of the machine was that the atomic particles were accelerated by electrical and magnetic forces to speeds which had never been reached before. The alpha particles from a radioactive source could then be given sufficient energy to enter the nucleus of atoms and as a result break them up. Fig. 13 shows an outline of the apparatus which is described later in Chapter III for the separation of uranium isotopes.

Fig. 13 Mass Spectograph (diagrammatic only)

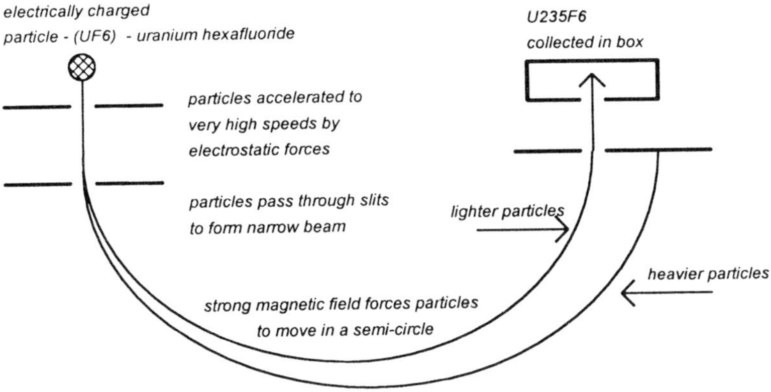

Magnetic field is perpendicular to drawing

Up to this time the method of investigating the particles produced by this bombardment had been to count visually the "scintillations" produced when they struck a specially prepared screen. As the counting was done by the human eye and brain it was not very reliable and counts of up to about 150 per minute was the limit. However, Hans Geiger, a German chemist, had already devised an electrical counter. This we now know as a "Geiger counter" which I am sure you have heard of, used for monitoring radioactivity. It consists of a wire sealed into a glass tube containing one of the rare gases - xenon - at very low pressure. The wire is charged electrically to a few hundred volts. When an alpha or beta particle enters the tube it ionises the gas, makes it conduct and the charge on the wire is altered. This alteration in charge is measured and after amplification can be heard as a typical "click". (The faster the clicks, the greater the radioactivity). or it can be shown on a television tube or a permanent record made for future study. This made the investigation of the particles produced much simpler.

The best source of alpha particles at the time was from polonium. This was difficult to obtain but Mme Curie was able to separate it from isotopes produced from the radioactive decay of radon (the radon was used in capsules for the treatment of cancer). Frederic and Irene Joliot-Curie bombarded beryllium (a metal similar to aluminium but even lighter) with these alpha particles and discovered a remarkable effect. When a substance such as paraffin wax which has many hydrogen atoms in its composition was put between the beryllium and the counter the rate of count (the clicks) increased dramatically. They reported this in January 1932 but erroneously suggested that the effect was caused by gamma radiation knocking a hydrogen atom out of the wax.

James Chadwick had been thinking about the neutron. He repeated Joliot-Curie's experiment and measured the range of the particles knocked out of the wax to be some 40 cm in the air and concluded that they were protons (hydrogen nuclei). He then went on to place other substances in place of the wax between what appeared to be the source of something being emitted from the beryllium. In every case the number of clicks increased. He deduced from this that the radiation from the beryllium could not be gamma radiation as suggested by the Curies as it simply would not have enough energy to produce the hydrogen ions from *heavier* and more complex atoms. If, however, instead of radiation they were particles of mass nearly equal to the proton and with no nett charge, the anomalies could be resolved. This then was the neutron which Rutherford had suggested in 1920.

Finding a neutral particle of similar mass to the proton we now had a means of getting past the electrical charges on the nucleus and as a result breaking it up with the results which we shall see in the next chapter.

CHAPTER III

FISSION AND PRODUCTION OF THE BOMB

In the September of 1939 the "Dogs of War" were released for the second time in the twentieth century and the struggle for survival was on.

Hitler, with his obsessive hatred of the Jews was not only persecuting and eventually exterminating them but causing the most gifted scientists, many of them Jews, to flee the country and settle in the West. For the Free World this was probably the most valuable thing that could have happened. Many of them were physicists and among them was a young Hungarian Jew - Leo Szilard.

In 1934, Marie Curie, the discoverer of radium, was mortally ill with leukaemia induced by many years of exposure to the radiation which, as mentioned in Chapter I, she may not have realised to be lethal. She lived however to see her daughter Irene and her husband Joliot make a discovery of the greatest importance in the development of nuclear reactions. They repeated the experiments which they had carried out and which were described in the last chapter, in which they had just missed the discovery of the neutron. They then found that on bombarding thin foils made of aluminium with alpha particles they had produced a type of phosphorus which was *radioactive* and had a half life of three minutes. This was remarkable, they had produced artificially a radioactive element. It meant that not only had they knocked a piece off the nucleus of the aluminium but in doing so **energy had been released.** This may have been heralded as "one of the most important discoveries of the century" but the Joliot Curies foresaw that, not only was there an enormous liberation of energy which might well be usefully employed, but also the possibility of it becoming an explosion and a cataclysm if the contagion spread to all the elements of our planet.

Up till then it had been considered that the nucleus was composed of single particles - the protons - but if this was the case neutrons would probably shoot straight through. In 1936 Niels Bohr again made another of his inspired suppositions. The nucleus of an atom could be made of neutrons and protons closely packed together so that an approaching neutron would be more likely to collide with it. This concept is now considered to be correct. (See Fig. 12 - the Helium Atom in the last chapter.) If you think about it, you may well ask what about the force of repulsion which must be present if we have a number of positively charged protons packed close together. Remember, like charges repel. The explanation is that we introduce another force which is called the "nuclear strong force" which I hoped not to have brought in so soon and which I shall discuss later. For the moment you can consider it as the "glue" which holds the nucleus together.

Leo Szilard had heard of the Joliot-Curie experiments and was convinced that it was the *neutrons* produced by the action of the alpha particles on the beryllium in their earlier experiments which were responsible for the break up of the nucleus. He also realised that if an element has its nucleus broken up by collision with one neutron and *two* or more neutrons are emitted as a result, then it was possible that a "chain reaction" might ensue with more energy being produced than was supplied by the initial neutrons. Where this energy could come from will be described later but if uncontrolled the reaction would produce prodigious amounts of energy - perhaps an atomic bomb! - and if it could be controlled, a source of power on an industrial scale. The problem was to find such an element and to harness the chain reaction so that the world would not be blown apart.

Following the Joliot-Curie experiments, it fell to Enrico Fermi, an Italian physicist, to carry on the work of bombarding other elements with neutrons. Instead of using polonium as his source of alpha particles he used the gas radon which was a much stronger source. Having worked through the elements, a massive task in itself, he came to uranium which up to then had been the heaviest element known and with the highest atomic number. To his very great surprise he found that some of the products from this bombardment must consist of elements of higher atomic number than uranium, hitherto unknown.

There were however some anomalous results which varied if his experiments were carried out on the normal wooden benches or on marble ones. Italy you may know is famous for the marble from Cerrara in the Apuan Alps in Tuscany and some of the benches in Fermi's laboratory had been made from this marble as it was more durable than wood. It was the fact that there were many hydrogen atoms in the molecules of wood which are not present in the marble (consisting of calcium carbonate = calcium, carbon and oxygen) which had caused these surprising results. Fermi then repeated what the Curie's had done, imposing a slab of paraffin wax between the neutron source (beryllium) and the target. He found, as they had done, a very strong increase in activity from the target. The conclusion he came to, which is fundamental to the understanding of nuclear reactions in bombs and power stations, is that the neutrons were *slowed* down by collisions with the hydrogen nuclei in the wax giving them more time to react with the nuclei of the target. Perhaps the Tuscan mountains helped in the development of the atomic bomb!

It has been calculated that a neutron at its normal speed would pass through a nucleus in less than a millionth of a trillionth of a second - if you can imagine how short a time that is - one second divided by ten with twenty-three noughts after it! So if one can slow the neutron down by causing it to collide with light nuclei such as hydrogen there should be a much greater chance of it breaking up a nucleus of a heavier element. Water was tried in place of paraffin wax and had the same effect as did some other lighter elements which led to some of the "moderators" used in our nuclear reactors and bombs.

The Liquid Drop Analogy

The next scientists to carry on the investigation were Otto Harn, a German physicist, and Lise Meitner, an Austrian Jewess. Harn continued work on bombarding uranium with neutrons. He found to his astonishment that the element barium - atomic number 56, some 36 places down the periodic table - was produced. Up to then only single particles had been chopped off the nucleus and it did not seem possible that the uranium nucleus had been cleaved right through to produce barium.

If we consider a drop of liquid such as water, it is held together as a drop by the forces of surface tension. This force is exhibited in everyday life. If you take two plates of glass, place a small quantity of water between them and try to pull them apart without sliding one over the other you will find it very difficult to separate them due to the force of surface tension. Considering the analogy of the water drop, Meitner realised that in the heavy uranium nucleus the mutual repulsion of the protons within it would work against the forces holding the "skin" together. It would become unstable just like a water drop which assumes a dumbbell shape when you shake it and only a small amount of energy is required to break it into two (Fig. 15).

Fig. 15 The Liquid Drop Analogy

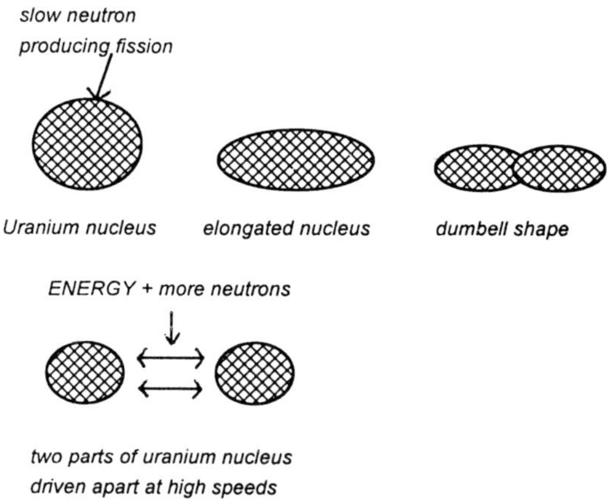

This would mean that even a small amount of energy from say a slow moving neutron could cause the uranium nucleus to split. This would explain the barium result. When this happened, the two parts of the split nucleus would be driven apart by their mutual repulsion at very high speeds which would mean that great energy was involved. Where did this energy come from? That was the question. Then Meitner remembered from previous work that when a nucleus splits into two smaller nuclei these would weigh less in total than the nucleus from which they had come. You may remember from the periodic table described in Chapter I that the atomic weight of hydrogen is not 1 but 1.008 and that helium is made up of four hydrogen nuclei or protons. However the helium atom does not weigh 4 time the hydrogen atom, which would be 1.008 multiplied by four = 4.032, but weighs (or more correctly has a mass of) 4.003. Some mass had disappeared. That was what was happening in the uranium nucleus. She also remembered Einstein's theory from which mass and energy are interchangeable and that energy equals mass multiplied by the speed of light squared (multiplied by itself) - an enormous number. She found that when she had worked it out, the energy calculated as that produced by the loss of mass agreed with the energy she and Harn had found.

This question of the correlation of mass and energy is fundamental to the whole question of atomic energy and the bomb. If matter disappears (and we have always thought this to be impossible) energy is produced. The equation which we have probably all seen is $E = M \times C^2$ where M is the mass, E is the energy and C is the velocity of light. As the square of the velocity of light is such an enormous number even a minute amount of matter can produce a prodigious amount of energy.

Otto Frisch was discussing these results with a friend who was a biochemist and compared it with the process in biological evolution where one bacterium divides into two and is known as "binary fission". As a result the term "fission" for the splitting of the uranium nucleus has become part of our language.

We have already noted that if the parts of the uranium nucleus when it splits could also produce more neutrons a chain reaction would ensue - more neutrons would collide with the nuclei and more energy would be emitted and so on. Fermi had carried out experiments which showed that indeed more than one neutron was emitted in this reaction so the way was now open for chain reactions to be considered as practical for the production of energy or a bomb.

In 1936, Leo Szilard who was still concerned with the political and social problems concerning the liberation of nuclear energy, proposed to take out patents on the subject in an effort to protect it from being exploited by unscrupulous people. Having been expelled from Germany he knew only too well the international situation and felt that the contents of the patents should be kept secret. He accordingly offered them to the War Office who declined them on the grounds that "there was no need for secrecy"! Szilard had been in contact with Professor F. A. Lindemann (later Lord Cherwell) head of the Clarendon laboratories at Oxford in the early thirties. Lindemann was an amazing character - the antithesis of what one would expect an Oxford professor to look like - immaculate in dark suit, bowler hat, with rolled umbrella and generally driving a Rolls Royce or Mercedes. He was a very rich man and a friend of Winston Churchill to whom he became Chief Scientific Advisor. Although Lindemann, like Rutherford, Einstein, and Bohr knew that the liberation of nuclear energy was possible none of them considered the odds on it being a practical proposition in the foreseeable future to be other than very long. However in spite of this he did write to the Admiralty on Szilard's behalf and they did agree that the patents should be kept secret and perhaps this may have had a part in hindering the German development of the bomb.

1939

In 1939, war was becoming inevitable and the scientists realised that the possibility of an atomic bomb had become a probability. Whether it could he developed in time to affect the war was still doubtful. More importantly, whether Germany would get there first was the question. The West had a start but at that time it was thought that the enemy were working on it. Could it be done in time to influence the war? Many people thought it impossible but war makes many "impossible" things possible.

In order to increase the probability of the neutron being captured causing the nucleus to split, it had already been shown that the slower the neutron - and here we are talking of prodigious speeds - the greater the probability of capture and fission. At the time the best way of slowing the neutrons was to pass them through deuterium, an isotope of hydrogen with a neutron in its nucleus making it twice as heavy as normal hydrogen and when combined with oxygen forms what is called "heavy water". (See Fig. 21 in Chapter IV)

Deuterium, and therefore heavy water, is present in ordinary water to a very minute amount and can be separated from it by a laborious and expensive process requiring vast amounts of electrical energy.

In 1939 it was known that the Germans were working on atomic bomb developments for which they required supplies of heavy water. They had control of the suppliers of this product - the Norsk Hydro plant in Vermork in Norway. The Allies decided that this plant must be destroyed. On the night of November 19th 1942, two gliders carrying parties of Royal Engineers and piloted by Army and Royal Australian Air Force glider pilots crossed the North Sea from

Scotland, towed by Halifax bombers of the RAF in appalling weather conditions. Both gliders and one bomber crashed and the survivors were executed cruelly on Hitler's orders. A second operation was mounted in February 1944 when six men - Norwegians, native to the region, parachuted onto a frozen lake north-west of Vermork. In spite of great difficulties and the hardest winter for many years, the operation was successful and the plant so heavily damaged that the work on the atomic bomb by the Germans was severely hindered. It is interesting to note that in February 1944 when the plant had been brought back into service the stocks of heavy water which were being transported to Germany were sunk in a lake in another covert operation. So Germany did not get the heavy water and we now know that work on the bomb never really got under way.

The Two Routes to the Bomb

Thorium, which you will have met with in Chapter I under radioactivity, is slightly radioactive but differs from uranium when it is bombarded with neutrons. Whilst both uranium and thorium both split (fission) when subjected to fast neutrons (those with greater energy), when the neutrons were slowed down by passing through paraffin wax, only uranium fissioned. This abnormal behaviour, that only uranium fissioned under the impact of slow neutrons could not be explained at first. Perhaps I am making it too complicated but I want to explain why uranium was the most suitable material for the atomic bomb.

Niels Bohr provided the solution. Natural uranium consists of the isotopes U 238 and U 235. Thorium has only one isotope Th 232. The U 235 is only present to the extent of 0.7% in natural uranium. Whilst the faster and more energetic neutrons can fission both natural uranium and thorium, the slower ones, although not able to fission U 238 and Th 232 are able to fission the U 235. He also pointed out that when U 238 absorbed a neutron it became an isotope with a nucleus of odd mass number and Th 232 became Th 233, also an odd number. On the other hand, U 235 became an isotope with a nucleus of even mass number. This made all the difference as it was later shown that if the nucleus of an atom changed from one with an odd number of neutrons to one with an even number, it produced considerably more energy. It was then apparent that U 235 had a great advantage over U 238 or Th 232 when it fissioned on account of the amount of energy produced. So if U 235 could be isolated this would form the basis of an extremely powerful explosive and was one route to the bomb.

When Enrico Fermi had bombarded other elements in his laboratory, which you remember was fitted with the marble and wooden benches, he had found, in the case of uranium, products with higher atomic numbers than uranium, which up to then were unknown. This work was continued by an American - Glen Seaborg - who had produced neptunium with At. No. 93 which was the next element in the periodic table above uranium. He was now looking for element 94. After a very complicated chemical process he succeeded in separating a millionth of a gram of neptunium which then decayed spontaneously to element No. 94 which was named "plutonium". Plutonium was then found to fission under the impact of neutrons as readily as U 235, and further, it could be made from natural uranium by absorption of neutrons in what came to be called an "Atomic Pile" which will be described in Chapter V. As it was not an isotope it could be separated by chemical means although with great difficulty. You will remember that chemical separation was not possible in the case of U 235 which had identical properties with U 238.

This then was a second route to the bomb.

The Preparation of the Bomb

"Manhattan" was the code name for the mammoth industrial project carried out by the American scientists and industrialists, assisted by their British counterparts, to manufacture the materials for, and to construct, an atomic bomb. The motivation was to produce a bomb, the possession of which, it was predicted, would end the war and save thousands of lives. The urgency was to achieve this before the enemy had a similar weapon. After the war Intelligence has shown that neither the Germans, the Japanese nor the Russians had progressed as far as the Allies in the production of a bomb but, not knowing this at the time, the pressure was on.

There were, as we have shown, two ways to approach the making of an atomic bomb - to separate the rare isotope uranium 235 from natural uranium or to make plutonium Pu 94 from uranium in an atomic pile. In 1943 the Americans decided to go ahead by both routes.

The Separation Processes

Uranium 235 is present only to the extent of 0.7% or one part in 140 of natural uranium. The two isotopes U 235 and U 238 (which make up natural uranium) are chemically identical which means they cannot be separated by chemical means.

Let me try to explain this: if you have a mixture of, for example, Washing Soda (sodium carbonate) and common salt (sodium chloride) which are chemically different, you could separate them by adding a weak acid such as vinegar (acetic acid). This would react with the washing soda but not with the salt and could lead to a means of separation. (I am not saying this would be a practicable method but it illustrates the principle). On the other hand if you have a liquid such as milk which consists of cream and skim, they are chemically similar but have differing physical properties and can be separated, taking advantage of their differing weights, either by leaving the milk in a dish to separate by gravity or by means of a cream separator (centrifuge) which substitutes for gravity a much stronger centrifugal force. These are very crude examples, and when you come to the uranium isotopes their physical properties, i.e. their atomic weights are extremely close together, making separation very difficult.

There were four ways which the Manhattan project considered for the separation of the isotopes:

(1) Gaseous diffusion.
(2) Liquid thermal diffusion.
(3) Continuous centrifuges.
(4) Electromagnetic separation.

Gaseous Diffusion

The principle of gaseous diffusion is that if you pass a gas through a porous material, for example china clay, the material from which clay pipes, well loved by "Popeye the Sailor", were made, the lighter component of the gas passes through more rapidly than the heavier. In the case of the uranium isotopes the difference is so small that to achieve a worthwhile separation requires the gas to be passed through the porous material very many times. The eventual plant for this process contained thousands of tanks and was of monumental size covering over 42 acres under one roof and four stories high. In order to make the uranium into a gaseous form, which was necessary to enable it to pass through the porous material, it was converted to uranium hexafluoride. This required large amounts of fluorine gas which up to that time had not been produced in industrial quantities. The hexafluoride was a very corrosive substance, the slightest contamination with grease or dirt could cause it to react violently. In order to move it about in the plant required pumps. Pumps normally have glands. The gland is the part of the pump which prevents the liquid or gas from escaping along the rotating shaft. These glands would have to be lubricated with grease which would have reacted violently with the hexafluoride. New glands with new materials or new types of pump had to be found to overcome this difficulty. One of the "spin offs" from atomic development was the invention of "Teflon" or PTFE - the material which is applied to your skis to make them more slippery and for example in "non-stick" pans. Another method was to do away with the gland altogether and to drive the pump through a magnetic coupling. These methods made the pumping of the hexafluoride possible. Fig. 16 shows the basic details.

A more serious problem was to find a porous material which would stand up to "hex" (as it was now called). China clay was no good. Nickel deposited by electrolysis was tried. Somewhat surprisingly, an interior decorator, Edward Norris, had patented a new kind of paint sprayer and this patent, adapted by Edmund Adler, produced a suitable porous barrier known as the Norris-Adler barrier, which with various modifications was used. The gaseous diffusion plant did eventually produce sufficient uranium 235 to allow the production of the bombs.

Fig. 16 Gaseous Diffusion

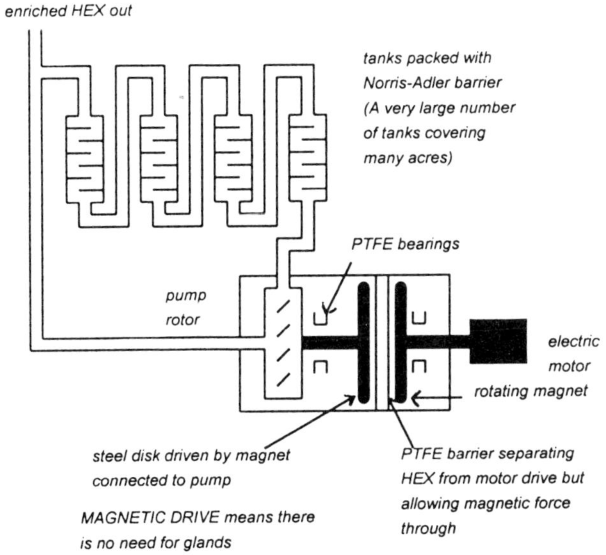

Liquid Thermal Diffusion

The principle behind this method is that in a hot liquid the lighter isotope tends to diffuse towards the hotter region whilst the heavier one diffuses to a colder region. Philip Abelson developed this method by arranging a hot pipe inside a cold one with liquid "hex" flowing between them. Not only would thermal convection - a hotter liquid is lighter than a cold one and therefore rises - cause liquid to flow up the surface of the hot pipe and down the cold one, but the U 235 enriched hex would come to the top of the pipe where it could be tapped off and recirculated round another series of pipes. With several thousand pipes Abelson hoped to produce 90% U 235 at the rate of one kilogram per day. In practice this was too optimistic and as it would require prodigious amounts of steam the Manhattan project did not proceed with it at first. However it could produces slightly enriched uranium (uranium with more U 235 than the 0.7% originally present) in quantity fairly rapidly and this was of use in the atomic piles used for plutonium production which will be described later.

Continuous Centrifuges

This method uses the same principle as the cream separator. Uranium hexafluoride is spun at high speed in a vessel where the heavier U 238 molecules move towards the outer wall whilst the concentration of U 235 molecules increases towards the centre. The enriched "hex" can then be drawn off and passed to another centrifuge for further enrichment and so on in a whole battery of centrifuges. Theoretically any degree of enrichment can be obtained but in practice only 3 or 4% was obtained. Fig. 17 gives an outline of one centrifuge.

Fig. 17 *Gas Centrifuge*

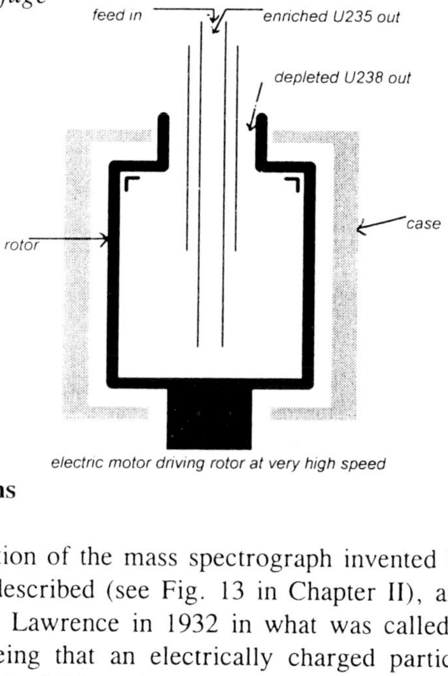

Electromagnetic Separations

This is really an elaboration of the mass spectrograph invented by Aston in 1918 and already described (see Fig. 13 in Chapter II), and further developed by Ernest Lawrence in 1932 in what was called a Cyclotron. The principle being that an electrically charged particle travelling through a magnetic field tends to move in a circle. The radius of this circle varies with the mass of the particle. Here then was a method of separating the U 235 from the U 238 atoms on account of their differing masses. Uranium hexafluoride in the form of a vapour is subjected to an electrostatic force which accelerates its molecules through a vessel. This vessel has had all the air and gases removed from it to reduce the number of collisions with "hex" molecules by pumping to a very high degree of vacuum. The molecules which are accelerated through it are deposited on a collecting plate at the far end. During the passage through the vessel, the "hex" is subjected to an extremely strong magnetic field. This causes the "hex" molecules to move in a semi-circle before being deposited on the collecting plate - remember that although the uranium is combined with fluorine to make "hex", the molecules of "hex" containing the two isotopes also have the slightly differing weights. The U 235 "hex" molecules which have described a semicircle will be deposited on the plate at a slightly different position than the U 238 and can therefore be separated. With a four foot diameter semicircle the separation was of the order of 0.3 inches.

To scale the plant up to make usable quantities of U 235 was a monumental task. Four foot vacuum tanks were set up between the pole pieces of large electromagnets. This arrangement, known as a "calutron" was multiplied many times and set up in the form of an oval like a racetrack and "racetracks" they were called (Fig. 18). The building to house them and the enormous pumps required to produce the vacuum covered an area which could have contained 24 football pitches. The electromagnets had to be wound with copper strips to carry the heavy electric current required. There was however a shortage of copper at that time and the best alternative would be silver. In order to provide the silver, the US Treasury agreed to release their stocks of silver bullion which were then melted down, cast into strips and used in the calutrons. The silver was eventually returned to the Treasury after the war, having played its part!

The magnets had to be cooled with circulating oil which in turn was cooled by water circulated round large cooling towers. With five "racetracks" it was estimated about half a kilogram (say one pound) of U 235 could be produced per month.

Fig. 18 *Calutron (racetrack)*

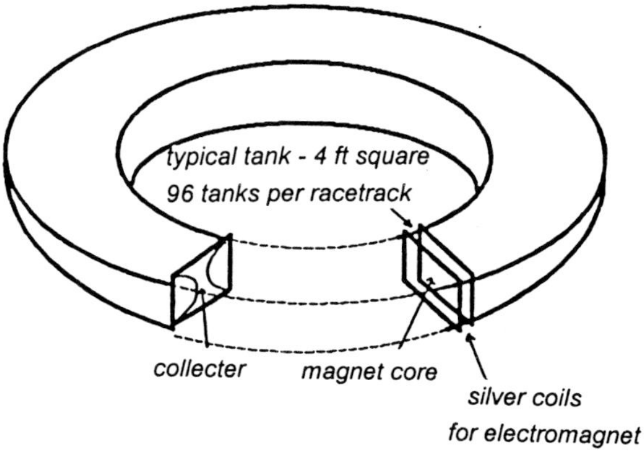

Plutonium

You will remember, when describing the second route to the bomb, that Enrico Fermi, who had carried out his experiments in Italy, had bombarded uranium with neutrons and had found that elements with higher atomic numbers, which had never been found before, wore produced.

The next higher element from uranium in the periodic table was neptunium - No. 93 which decayed naturally in a short time to the next element No. 94 which had been named plutonium.

Plutonium fissioned as readily as uranium 235 when struck by neutrons and so could be used in a bomb. Its preparation does not have the same problems as the separation of the uranium isotopes and it could be produced in an atomic pile fed with natural uranium.

The Manhattan project therefore proposed building an enormous pile with the sole purpose of producing plutonium.

The pile, which we can consider as the forerunner of atomic piles used for power generation, consisted of uranium enriched with a higher proportion of U 235 (see Liquid Thermal Diffusion) and moderated with graphite to slow the neutrons down so that fission could take place. To control the pile, rods of cadmium which is a strong absorber of neutrons were inserted, and these would reduce the rate of fission of the uranium.

The heat generated in the pile which could have been used for power generation was wasted and removed by cooling water.

The original piles which were constructed at Hanford near the Columbia river were to be cooled by helium under high pressure. This was an expensive option and it was later found that, if the uranium metal was encased in aluminium cans and very pure water was used for cooling, this was a satisfactory alternative. Eventually three piles were built each producing the equivalent of some 250 megawatts of power. This power in the form of heat was removed by cooling water from the Columbia river raising its temperature by several degrees!

After about 100 days approximately one atom in 4000 of uranium had been converted into plutonium. These "slugs" of uranium containing this minute proportion of the man-made element, still enclosed in their aluminium cans, were dropped into a a deep tank containing pure water.

Apart from the plutonium which is only mildly radioactive but highly poisonous if ingested, some products were produced which had very strong radioactivity but fortunately short lives.

Due to their strong radioactivity, these short-lived products produced an unearthly bluish glow in the water surrounding them. You may have seen this in the James Bond films when the villain is about to be pushed into a tank of highly lethal radioactive liquid!

After 60 days this radioactivity had decayed and the slugs could be removed and sent for chemical separation of the plutonium.

When you remember that only microscopic amounts of plutonium were made in the original process, this was an amazingly successful operation.

Critical Mass

You may have heard of the "critical mass" in connection with the atomic bomb. This refers to the amount of uranium which would explode spontaneously. It is based on the fact that the surface area of a sphere increases much more slowly with increasing diameter than does the volume.

Some of the free neutrons which are moving about in the natural uranium do escape from its surface. At a particular volume more neutrons would find nuclei to fission in the uranium than would escape from the surface and a chain reaction would ensue.

This critical mass was difficult to calculate and at first was put at forty-four tons - an enormous amount of natural uranium metal and unlikely to have been achieved or the world would have been blown up long ago!

Later, with uranium enriched with its U 235 isotope, this calculation reduced it to the size of a coconut and, for pure U 235, the size of a golf ball!

No one particularly wanted to try it, for fairly obvious reasons, but eventually the following experiment was carried out.

A slug of uranium was allowed to fall by gravity through a hole in a piece of enriched uranium and the instantaneous concentration of neutrons measured. This confirmed the calculations that a reasonable sized piece of U 235 would produce a bomb.

In 1940 it was apparent to Otto Frisch and Rudolph Pearls that a bomb of immense destructive power was possible and the only defence against it was the counter threat of a similar weapon. It may be argued - why did they not forget the whole idea when they realised it would produce a weapon of unparalleled violence? The answer was - we were at war.

The Bomb

The Manhattan project, as we have seen, was set up to produce an atomic device which would have an explosive effect of tremendous proportions. It was hoped that the Allies would attain this object before the enemy and that the knowledge that they had such a weapon and had proved that it worked, would end the war.

I have described how the scientists approached the task and how the two products, uranium 235 and plutonium, could be produced in sufficient quantities to produce such a device.

The first option was to use U 235. Two blocks of U 235, each less than the critical mass required for a chain reaction to proceed, but together equal to this mass, had to be brought together so fast that there would be no time - and here we are talking of millionths of a second - for them to disintegrate before the devastating chain reaction evolved. The device which was named "Little Boy" was essentially a gun closed at one end by a ring of U 235 imbedded in a steel block or "tamper" into which the second block of U 235 was fired by an explosive charge. Just imagine a field gun being fired with its muzzle welded up! However it worked (Fig. 19). Little Boy was two and a half feet long, over two feet in diameter and weighed over four tons. This was the device which was dropped on Hiroshima from the United States B29 bomber - "Enola Gay" - on August 6th 1945 with the devastating results which have been described so often I will not try to cover them again.

Fig. 19 'Little Boy' (Uranium Bomb)

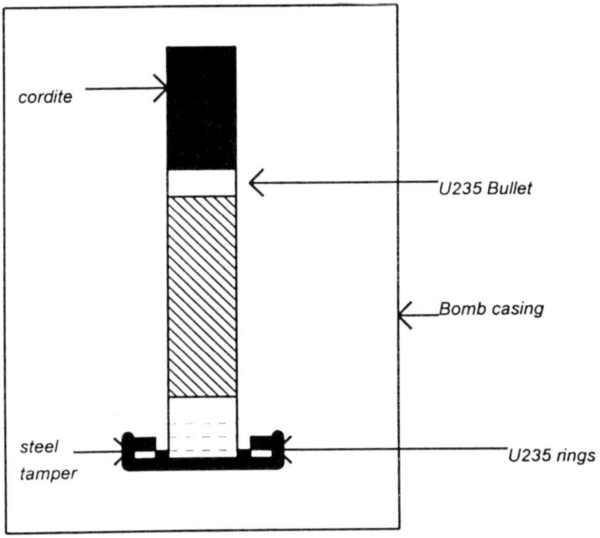

The second option involved the use of plutonium, and although this was slightly less difficult to produce than U 235, several problems arose. There is another isotope of plutonium - Pu 240 - which is produced at the same time in the atomic pile used for producing Pu 239. This isotope had a much higher rate of spontaneous fission than the Pu 239. This meant that with a gun, as used in Little Boy, even with a speed of 2000 ft per second, the plutonium bullet would not be travelling fast enough to cause complete fission of all the plutonium before the mass would melt and the bomb would "fizzle" (this was the word used). Another method had to be devised.

After many suggestions had been considered it was decided to proceed as follows. If a block of plutonium slightly larger than its critical mass could be suddenly compressed so that its mass reached criticality, and if at the same time neutrons could be fired into it, an explosion would take place. The problem was how to do it. The solution was to make a plutonium core in the shape of a sphere and surround it with a number of explosive charges. The explosive wave from these charges was concentrated inwards by a method used in conventional armour piercing projectiles. The charges were shaped like the reflectors of your motor car headlights but instead of parallel beams they concentrated each beam to a point with the result that they could punch a hole in the armour of a tank. In the case of the plutonium bomb, this caused an "implosion" as opposed to an explosion which compressed the plutonium to its critical mass, when neutrons would set off the explosion. There were two other important requirements. The first was an initiator to produce the stream of neutrons at the vital moment when criticality had been reached. A polonium/beryllium device was used but this aspect is, as far as I know, still secret. The second was that all the explosive charges placed round the periphery of the sphere with the plutonium at its centre must be fired instantaneously, which means, in this case, within a fraction of a millionth of a second. The electronic device required to fire all the charges simultaneously is known as a "Krytron" or "Sprytron" and is capable of delivering a pulse of electricity of 200 amps. at 2000 volts in less than half a microsecond (remember one microsecond is one millionth part of a second).

The detonators are like an ordinary fuse wire in your house or motor car. When subjected to the large electrical pulse from the Krytrons they vaporise and start the explosions. It was these Krytrons which set off such a furore when it was found they were being exported to Iraq who would have needed these miniaturised devices to set off the detonators in the long range guns they were presumed to be making, or in an atomic device. Fig. 20 shows in outline what a plutonium bomb might look like.

Fig. 20 'Fat Man' (Plutonium Bomb)

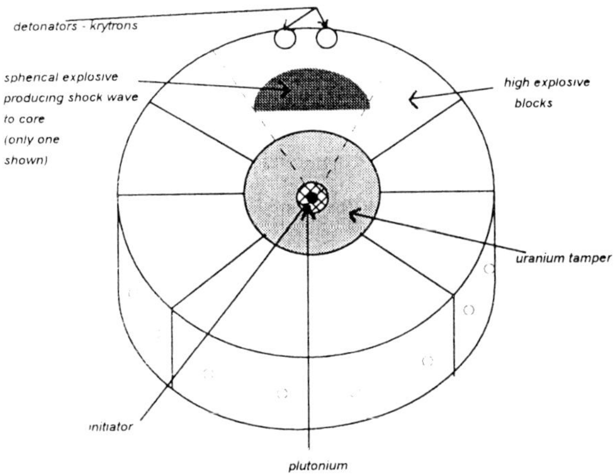

Whilst it was considered that Little Boy would work without the need for an experimental firing, "Fat Man" - which was the name allotted to the plutonium bomb, was quite different, and it was agreed that a test explosion should be carried out before any bombs were dropped on Japan. A site for the explosion of the first nuclear device was chosen in the desert of New Mexico, 60 miles north west of Almagordo - known in Spanish times, perhaps appropriately as Dead Man's Trail or the Journey of Death. The device was to be fired from a steel tower 100 feet high. The results were recorded on high speed cameras as well as being viewed by personnel who were sited some four and a half miles away. The core of Pu 239 weighed about 11 pounds and could not have been much larger than a cricket ball - the Americans probably said a baseball; so much energy in so small an object! In the early morning of July 18th 1945 the device was fired and the release of nuclear energy by the human race on a gigantic scale took place for the first time. On August 9th 1945 a B-29 bomber - named "Box Car" - took off from Trinian Island in the Pacific and at 11.02 Fat Man exploded 1650 feet above Nagasaki.

CHAPTER IV

FUSION, THE HYDROGEN BOMB, AND POWER BY FUSION

When we were discussing fission in the last chapter, the energy was produced by the loss of mass when the uranium nucleus split into two parts which weighed less in total than the original nucleus. This energy which appeared as heat and radiation in the atom bomb was the energy predicted by Einstein from his well known equation relating mass with energy.

If, instead of splitting the nucleus, we are able to bring two nuclei together to form another nucleus which weighs less than the two nuclei from which it is made, again energy will be produced as some mass has disappeared. The researchers into this subject considered that the best approach would be to consider the lightest atom - hydrogen. If four hydrogen nuclei could be made to combine together, or as we now say to "fuse", to form one helium nucleus some mass would disappear and energy would be produced. Remember one hydrogen atom (or its nucleus) weighs 1.008 units and one helium atom 4003 units so 0.029 units (4.032 - 4.003 = 0.029) must have vanished. The actual weight of the hydrogen nucleus is minute, so even multiplied by the speed of light squared the energy produced by the fusion of four hydrogen atoms is very small. But remember there are millions and millions of them so the possible production of energy is enormous.

However if we are to get the nuclei to come close enough together to fuse we must propel them with an enormous velocity to get them past the very strong force holding the nuclei together - the nuclear strong force which we shall come across in Chapter VIII. One way of doing this is by heating which causes the atoms to move faster. This would require temperatures of millions of degrees, far hotter than anything achieved at that time on the Earth but approaching the temperatures we now consider to be those at the centre of the Sun. It was then realised that the atom (fission) bomb produced temperatures of this order. Why not use an atom bomb to initiate the fusion of the hydrogen atoms and the energy produced would proliferate as long as there were hydrogen atoms present.

This then was the possibility - a horrendously frightening one when considered by the scientists in 1942 - the world might be vaporised and disappear into space! The reaction, unlike fission, did not require a critical mass and once started fed on hydrogen atoms - and there were enough of those in the world - consider the oceans alone! This cataclysmic scenario was checked and counterchecked and fortunately for our survival it was proved that under the most violent reaction conditions that could be envisaged, the loss of energy by heating the surroundings would always exceed the energy produced by a very large factor which would prevent the reaction running away and the world would not ignite.

The research continued but instead of trying to fuse four hydrogen nuclei, it was suggested that it would be easier to use two *deuterium* (hydrogen 2) nuclei. You will remember that deuterium or "heavy water" is present in ordinary water to a very small amount and was used as a moderator in the earlier work on fission.

The reaction would then involve two deuterium nuclei fusing to form a helium nucleus with the ejection of a neutron and a release of energy. A second reaction would form *tritium* (hydrogen 3). Tritium is another isotope of hydrogen with one proton and two neutrons which does not occur in nature but can be produced by bombarding lithium with neutrons. It is also radioactive. This then produces a further release of energy (Fig. 21).

Fig. 21 *Isotopes of Hydrogen*

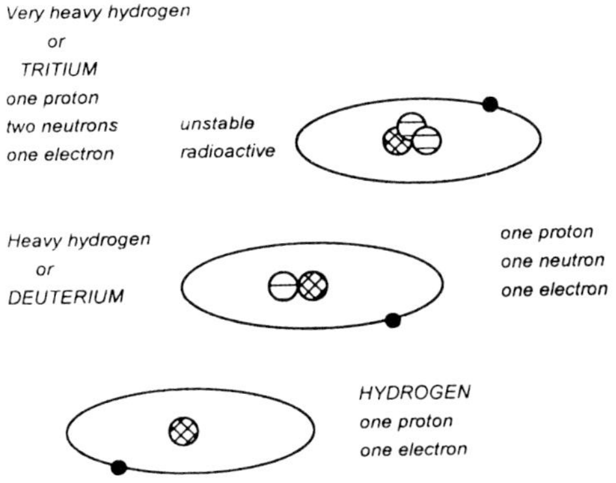

The birth of the H bomb has always been attributed to Edward Teller, one of the Hungarian Jews who left Germany before the war and emigrated to America where he became one of the leading scientists in the Manhattan project. However it appears he was forestalled by a Japanese physicist - Tokaturo Nagiwara at the University of Kyoto who in May 1941 considered using a nuclear chain reaction to institute a thermonuclear reaction in hydrogen. Perhaps it was as well that the Japanese did not proceed further with this development or history might have been rewritten.

The hydrogen bomb or "super" as the Manhattan project called it, would consist of a mass of liquid deuterium plus an amount of tritium to be exploded by an atomic (fission) bomb. This was in 1946. The problems were: first to produce the deuterium from water by electrolysis as has already been described in the previous chapter - this was an easier problem than producing U 235 from natural uranium - and then to liquefy it and keep it liquid. This required very low temperatures, lower than those required for liquefying air. Next, the tritium would have to be produced by bombarding lithium with neutrons - (lithium is a metal similar to sodium and could be produced without much difficulty). Finally the mathematical calculation of the sizes and proportion of the constituents required were prodigious. At that time computers were in their infancy but Stanislaw Ulan, a mathematician who came to America from Poland in the 1930s was able to carry out the essential but laborious calculations which made the bomb possible. If they were wrong, the bomb would melt or blow apart long before the pressure and temperature were sufficient to initiate the fusion reaction. (The machine which made these calculations possible was called "Mathematical Analyser Numerical Integrator and Computer" - MANIAC for short - a very appropriate acronym!)

The first experimental thermonuclear device was codenamed "Mike". It was fuelled with liquid deuterium and tritium requiring low temperature refrigerating equipment to keep them liquid and weighing some 65 tons. .It was sited on the small atoll of Elugelab in the Central Pacific near the island of Eniwetoc. It was fired on the 1st November 1952 and exploded with a force equivalent to 4 million tons of TNT - one thousand times more violent than Little Boy. The fireball extended to 3 miles in diameter, millions of gallons of water were turned to steam, the island of Elugelab vanished - vaporised - leaving a crater half a mile deep and two miles wide.

So the thermonuclear age had arrived.

I think it should be remembered that in 1934 Lord Rutherford had demonstrated in his cloud chamber that two hydrogen atoms had collided to form one helium atom and no doubt realised what it might mean. He would have been amazed but perhaps saddened if he had seen the result in 1952.

Mike was obviously too large at 65 tons to be feasible as a bomb to be delivered from an aircraft. A more suitable bomb was developed later. This used lithium deuteride which is a stable powder and is made by combining deuterium with lithium. The lithium used is an isotope (Li 6) which is present to the extent of 7.4% in natural lithium from which it can be separated comparatively easily. When the bomb was fired, neutrons from the fission component of the bomb would produce tritium instantly from the lithium 6 and the tritium would then fuse with the deuteride to develop the thermonuclear explosion. To concentrate the explosion of the fission device it was surrounded with a case of metallic uranium ground in the form of lenses to focus the energy onto the core as had been done with Little Boy. The method of initiation did not rely solely on the heat and pressure from the fission explosion but made use of the gamma rays produced at the same time. The speed with which these rays were propagated initiated the thermonuclear reaction before the "melt down" could take place and destroy the bomb. Fig. 22 shows an impression of how an H bomb might be constructed.

This device was tested in the spring of 1954 and produced an explosion equivalent to 15 million tons of TNT and was in a form suitable for delivery from an aircraft. On the 12th August 1953 the Russians exploded an H bomb in Siberia. In May 1957 the British tested a bomb on Christmas Island in the Pacific but it was claimed that it was not a true H bomb. They did, however, test a true H bomb in 1958.

The Hydrogen Bomb had arrived.

Fig. 22 Possible design for Hydrogen Bomb 'The Super'

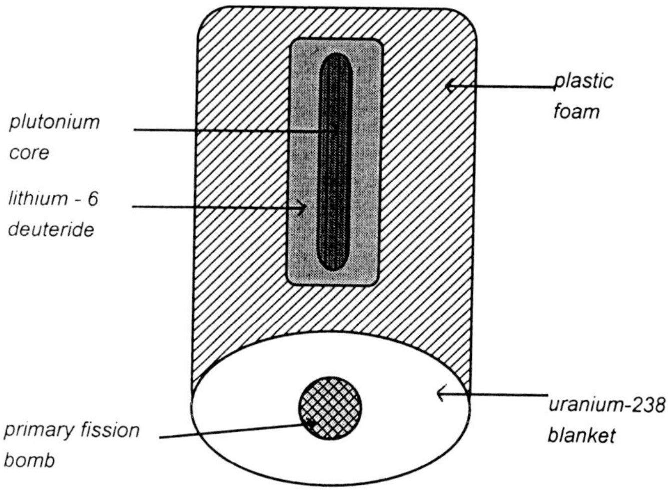

Power Generation by Fusion

Whilst the super was being worked on by the scientists of the Manhattan project it was obvious to them that here was a vast reservoir of power for the energy hungry world - if it could be harnessed. The energy theoretically available from one tenth of a gram of hydrogen to form helium was the prodigious amount of two and a half million kilowatts, sufficient to keep the electric lights burning in over a million homes. Not only was this an enormous source of power but it only emitted neutrons which could be shielded comparatively easily and there were no radioactive residues to be disposed of as in the present generation of nuclear power stations. So far the fusion process required a fission reaction to initiate it and there was no way of controlling it. To make a worthwhile proposition the gases would have to be raised to the enormous temperatures required but under controlled conditions. This then was the problem.

If by applying enormous electric power in short pulses of hundreds of megawatts (the output of an average power station) you can succeed in raising the temperature of the deuterium to that required for fusion - how do you contain it? No vessel nor any material can do this. The solution to this part of the problem is to contain the deuterium by means of a strong magnetic field so that it does not touch the walls of the containing vessel. The deuterium as well as being isolated from the walls of the vessel is said to be in the state of a "plasma". This means that it is completely ionised with equal numbers of positive and negative ions and is highly conducting. The plasma is contained in a "torus" - a vessel shaped like a "doughnut" and known as a "Tokamak" (Fig. 23).

The torus is surrounded by immensely powerful electromagnets which keep the plasma from touching the walls. The vessel is pumped to a very high vacuum and the plasma heated by the electrical pulses until the deuterium nuclei are moving at such high speeds that they overcome their inherent electrical repulsion and fuse. These machines cost billions of pounds. The most successful one at the moment is the "Joint European Torus" or "JET" which operates at the Culham Laboratory of the UK Atomic Energy Authority. Although it has been possible to produce fusion in the JET machine, the amount of energy released is far less than that required to power the machine and to get the "burn" started. So, sadly the hoped of limitless power is still in the future.

Fig. 23 Tokamak (Doughnut shaped Torus)

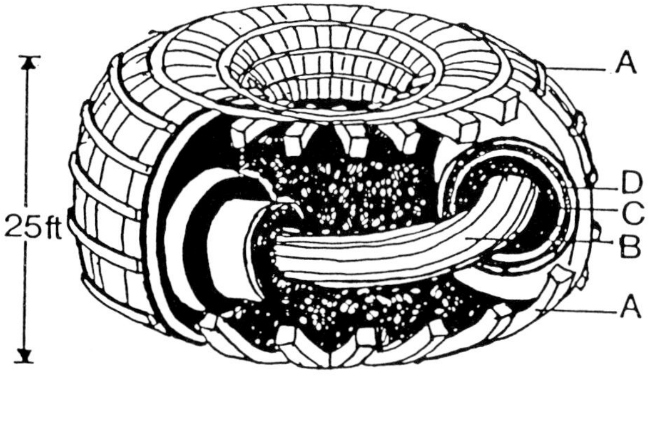

A - Electromagnets
B - Deuterium Plasma
C - Helium passing heat to turbines
D - Lithium blanket

A group of international scientists are planning to build an "International Thermonuclear Experimental Reactor" - ITER in Germany. This is intended to solve the practical problems as to how to extract usable quantities of energy greater than has had to be put in. It will use superconducting electromagnets (see Chapter VI) and various methods of heating including radio frequencies as used in your microwave oven. Sadly it will be many years before we see any results.

Another approach is to make use of a number of very powerful laser beams (Chapter VI) concentrated on a pellet of solidified deuterium and tritium. The sudden heat produced almost instantaneously produces a pressure wave which causes an implosion in the pellet. The heat and pressure is then sufficient for fusion to take place. Work on this is proceeding.

The idea of producing fusion without recourse to high temperatures was considered by a British physicist, Sir Charles Frank in 1947. His idea was to use a rather exotic sub-nuclear particle which we shall meet in Chapter VIII under "Cosmic Rays" called a *muon*. A muon is somewhat like an electron but 200 times heavier and only lives for some two millionths of a second. Muons are believed to play a vital part in the cohesion of the nucleus so if they could be introduced into the hydrogen nucleus, fusion might more readily be achieved. However, as far as I know this method has not been proceeded with.

In 1989 there were some remarkable claims by two American researchers that they had produced fusion without the need for the. high temperatures conventionally required. Basically their experiments were very simple. They passed a small electric current between platinum and palladium electrodes in a solution of lithium in deuterium. They claimed that the deuterium nuclei absorbed by the palladium were brought so close together that they fused producing energy. Their claim has however been flawed. If genuine fusion had been taking place some very high energy neutrons would be produced and these have not been found.

Before leaving fission, fusion and the bombs there was another weapon considered in 1978 called the "Enhanced Radiation Weapon" (ERW) or the neutron bomb. It produced low heat and blast but high energy short range neutrons lethal against personnel under battlefield conditions but did little damage to buildings. It has been described in certain quarters as the "Ultimate Capitalist's Weapon".

CHAPTER V

NUCLEAR POWER GENERATION

We have seen in the preceding chapters how the two atomic bombs were developed and produced during the Second World War and how the "fusion" or hydrogen bomb was being developed and the possibility, and it was only possibility, of using this fusion as a means of producing useful power.

At the end of the war, the priority for nuclear research was still for the production of material which could be used in the manufacture of atomic weapons, although the use of a controlled nuclear reaction for ship or submarine propulsion was also being actively considered. As plutonium had been shown to be a more efficient material for the manufacture of atomic weapons than uranium 235, and as stocks of uranium metal were likely to be restricted for various reasons, which included political ones, production of plutonium was considered to be the first priority.

The method used during the war for producing this material had involved constructing an atomic pile, using enriched uranium, and the heat produced being wasted. However a pile fuelled with *natural* uranium would not only produce the plutonium required but would also produce heat which could be made available for power generation. This then gave an added impetus to construct a nuclear fuelled power station and was no doubt why the early work on such a project was shrouded in the clothing of the Official Secrets Act.

As has been explained in Chapter III the nucleus of the uranium *235* atom does split spontaneously emitting a neutron and releasing energy as a result of the annihilation of some of its mass. Natural uranium, you may remember, contains only about 0.7% of the uranium 235 isotope and although this isotope is producing neutrons by "fission", they are travelling at such high speeds that they are not pulled into the nuclei of the more abundant U 238 and only strike those that are directly in their path. Remember that as the neutron has no electric charge it is only attracted to the nuclei by gravitational forces which are much weaker than the electrostatic ones. However if the neutrons are slowed down many more collisions will occur and the chain reaction can be initiated in the natural U 238. Apart from the production of energy, the result of this reaction, as has been described in Chapter III, is to produce neptunium (element 237) which decays spontaneously to plutonium which can then be separated.

To slow the neutrons down, a material known as a "moderator" is inserted in the pile. This reduces the neutron's speed by such an amount that with a certain mass of uranium (known as the "critical mass" - which we have met before in Chapter III) the chain reaction proceeds.

By removing the moderator or by inserting a material which absorbs neutrons the reaction can be controlled so that it does not get out of hand. The moderator normally used is graphite (a form of carbon - a component of the "lead" in a pencil) and the heat produced in the pile was removed by circulating water. However water also absorbs neutrons and slows the reaction so if the water supply should fail, more neutrons would be available for fission and the pile might go out of control. Cooling by air was considered to be safer as even if the pumps circulating the air should fail, the pile was not relying on the circulating water to absorb some of the neutrons. As a further precaution neutron absorbing rods could be inserted rapidly thus shutting the reactor down. To guard against the possibility of mechanical failure, the rods could be dropped into the core by gravity. Air was therefore chosen as the preferred cooling medium although it was less efficient and the pile had to be run cooler which meant that the turbines used for the conversion of the heat into electricity were less efficient as they had to make use of low temperature steam. It is a fact, based on the laws of thermodynamics which I will not go into, that a turbine converts the heat from the steam driving it more efficiently if the steam is at a high temperature and pressure.

The uranium fuel elements were at first enclosed in aluminium cans which were then inserted into the graphite blocks through which the cooling air flowed. The cans were essential to prevent any radioactive fission products from being carried into the air stream and possibly into the atmosphere. It must be remembered that radiation is not carried by the air itself, if it is pure, but only by any irradiated dust or radioactive elements which may escape into the air stream.

Careful monitoring of the air passing to the atmosphere is an essential part of any nuclear reactor.

Another problem which had to be considered was that graphite when irradiated becomes distorted (the "Wigner" effect - called after the Hungarian physicist, Eugene Wigner, who first discovered it). This meant careful design, and operation of the plant to minimise this effect, which could be overcome by periodic annealing of the graphite. (See Windscale Fire.)

Windscale

The first reactor to commence working was at Sellafield (Windscale) in October 1950, and was a simple air cooled reactor, the air being discharged into the atmosphere and no power being generated, the object being to produce plutonium which was recovered from the fuel elements. Two units were built at Windscale but plans to build a third plant to produce plutonium were abandoned. The reason for this was because an isotope separation plant was started at Capenhurst in Cheshire, using the techniques already described for separating the U 235 isotope and thereby "enriching" the natural uranium fuel. This meant that the Windscale type of plant was not required apart from the production of plutonium.

Calder Hall - Magnox Power Stations

The first nuclear power station in this country was commissioned at Calder Hall close to Sellafield in Cumbria in October 1956. It used natural uranium as a fuel moderated with graphite and cooled by carbon dioxide, pressurised to increase the efficiency with which it transferred the heat from the pile to the steam producing equipment. The steam then drove the turbines at greater efficiency than in the original air cooled reactors as already explained. The fuel elements were of natural uranium metal rods approximately one yard long and one and a quarter inches in diameter.

They were encased in a can of magnesium alloy - "Magnox C" - which was both stronger and less neutron absorbing than the cans used at Windscale.

The name "Magnox" has been given to all natural uranium gas cooled reactors similar to the Calder Hall design.

There were altogether eleven Magnox reactors in the UK and they all produced plutonium which could be extracted from the fuel rods when they had completed their useful life. Calder Hall is still operating and licensed until 1996 - a life of forty years.

Hinckley Point "B" - AGR Reactors

The next reactor to be considered was the advanced gas cooled reactor "AGR". This used uranium oxide pellets, made from uranium enriched to 2 or 3% of U 235 from Capenhurst as the fuel and was cooled by carbon dioxide at a pressure of approximately 200 lb. per square inch. Much higher temperatures were achieved but the canning of the fuel elements at this temperature was a problem. Beryllium was found to be the best material for the purpose but it was expensive and was later superseded by stainless steel. The first station was opened at Hinckley Point and attained full power in October 1965. Since then 6 more AGR stations have been commissioned in the UK.

Dounray - FBR Reactor

Meanwhile experimental work was being carried out on a fast breeder reactor (FBR) at Dounray. This had a core of uranium 235 and plutonium from the spent fuel from other reactors, surrounded by a "breeder" blanket of natural uranium fuel elements. It could also use thorium as a breeder blanket which would be transmuted into another fissile isotope - U 233. Fast breeders would increase the U235 content of natural uranium from the existing 0.7% to 50% or more and would produce more plutonium than was originally present. The Dounray fast breeder reactor was operational in March 1955 and worked for seventeen years producing electricity and many radioactive materials for other commercial and medical uses. Its heat output was of the order of 600 megawatts per cubic metre of core against only 0.5 megawatts for the Calder Hall "Magnox" stations.

It was, as its name implies, a reactor using fast neutrons and did not require a moderator as in the other reactors. As a result it was necessary for stringent safety precautions be taken. Twelve of the fuel elements were designed to drop from the core in an emergency for rapid shut down and three boron neutron absorbing rods could also drop into the core. The core was cooled by a liquid sodium/potassium alloy with heat exchangers to transfer the heat into high temperature steam for the turbines. The whole of this complex equipment was enclosed in a special steel spherical containment vessel 32 metres in diameter and now a well known landmark in the Caithness coast.

Sodium and potassium , as those of you who have done chemistry at school will know are very reactive metals exploding with violence if a small lump is dropped into water or acid. I hope none of you will have carried out "experiments" which many years ago as a schoolboy I and other "confederates" carried out, after hours, in the chemistry laboratory.

A beaker with water or acid was placed in a sink and a piece of sodium or potassium placed on the bench beside it. Then, having taken cover, a projectile was propelled along the bench knocking the sodium into the beaker. We competed for the loudest "bang" and frightened the caretaker out of his wits!

Subsequently I was employed in the manufacture of sodium at a well known chemical firm. One of the characters working on the process, wishing to remove some rats which had infested the premises, waged war on them with lumps of sodium which effectively blew them to pieces! Having seen molten sodium metal being run into moulds and the catastrophic effect if any water is present, I fully realise the engineering problems in pumping liquid sodium in a reactor and then transferring its heat to circulating water. However, modern methods of construction have made this method perfectly safe.

The need for fast breeder reactors, such as that at Dounray, has been the subject of much controversy. Although the reactor at Dounray had operated for many years without danger, the public considered it to be more hazardous than other nuclear reactors. Its advantage was that it does not need a continuous supply of uranium and if uranium supplies were to be interrupted or limited then it would be of great value. As, however, it now seems that there will be adequate supplies of this material for the foreseeable future and as the requirement for plutonium is likely to be minimal, the need for FBRs has receded and Britain has pulled out of the project, leaving France and Germany to exploit it.

An enormous amount of money has been spent over the last thirty years on developing the FBR by Britain and its protagonists say it should not be wasted. It seems therefore that at least the results of the research should be kept as an insurance against future possible eventualities.

Pressurised Water Reactors - PWR
Sizewell "B"

Fig. 24 Pressurised Water Reactor

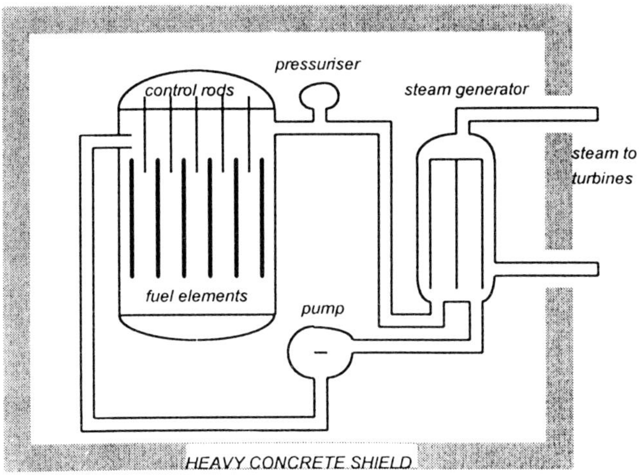

The next reactor to be looked at was the pressurised water reactor (PWR) which used uranium dioxide pellets as the fuel. It was considered for submarine propulsion for which nuclear power was a breakthrough as, not requiring oxygen, the vessel could remain under water for long periods with great tactical advantages. The PWR used the hydrogen (in normal - "light" - water) as moderator and coolant but this required the core to have a high level of enrichment to make up for the absorption of neutrons by the water. In order to prevent the water boiling it had to be kept under a pressure of nearly 2000 lb. per square inch - a fairly high pressure to be used under such working conditions where leaks containing radioactive material could be disastrous. The canning material used was zirconium, a scarce metal but the best for this purpose. The PWR was not at first considered suitable for land based nuclear power stations. However in view of its greater efficiency and reduced problems in waste management and disposal, it was adapted, but with the addition of very stringent safety precautions, for a new generation of nuclear power stations in the UK commencing with Sizewell "B". In France the use of PWR reactors accounts for some 70 percent of their electricity production (Fig. 24).

Steam Generating Heavy Water Reactor - SGHWR
Winfrith Heath

A variation of the AGR was the steam generating heavy water reactor - "SGHWR", which as its name suggests used heavy water (water with the hydrogen isotope deuterium - see Chapter IV - nuclear fusion) as a heat extracting medium. This was less neutron consuming and generated steam via heat exchangers. The coolant was at a temperature of 750 degrees centigrade and a pressure of about 300 lb. per square inch. Graphite was used as the moderator but as there was a possibility of radioactive materials being discharged to the atmosphere the exit gases were passed over charcoal cooled with liquid nitrogen. The reactor at Winfrith Heath was commissioned in 1968.

Hazards and Safety

The question of safety is paramount, and rightly so, in any discussion on the use of atomic power. The lack of knowledge and the fact that we cannot see or feel radiation makes anything to do with it most frightening.

As already pointed out, the air emitted from an atomic power station, so long as it is not contaminated (i.e. pure oxygen and nitrogen) is not radioactive, neither is the water (if it is pure). However, if any particles of dust or other materials which have been in contact with the core escape then this does constitute a serious hazard.

In the normal working of an nuclear power station and if, and some will say this is the big "if", all monitoring instruments are working efficiently and their results acted on at once there would be no danger. If again, or perhaps when, there is a major incident such as at Three Mile Island or Chernobyl the situation is very different. Apart from blast and heat in the immediate vicinity which, due to the containment vessels and walls, would be less than in the case of an atomic bomb explosion, the immediate effect would be a release of gamma radiation akin to a massive dose of X-rays. There would then be a distribution of debris and dust over a wide area and this would almost certainly be radioactive from contact with the core. As you will remember the radiation will consist of alpha, beta and gamma rays and probably neutrons. The first will be stopped by very little shielding or cover, as will the neutrons. The second by a little more but the gamma rays will be very penetrating. The additional problem is that a large number of possible radioactive isotopes may be produced with very varying lives (from seconds to many years) - some are absorbed and then secreted by the body and some are retained for a long time.

The problem which faces us at the moment, and which is not really satisfactorily determined, is what is the long term risk after exposure to radiation? It is generally accepted that a person who has been exposed to radiation , and I take it to mean gamma radiation, runs an extra risk of contracting cancer. However the strength of radiation concerned, the time of exposure, and how cumulative it is, are subjects of considerable controversy and statistical investigation. Although much has been written in the Press and technical journals, no clear and incontrovertible evidence has been found as to how great this risk is.

We must remember that we are all subject to radiation - cosmic rays (see Chapter VIII) flow through us all the time and cause no ill effects - the background radiation from rocks, especially granite, varies at different locations and does not appear to have serious effects. Recently there have been indications that the radon gas which issues from some rock formations in some locations is a health risk. We may also ask ourselves what about the enormous proliferation of electromagnetic radiation which is flowing through and all around us all the time?

It has been estimated that nearly 50% of the natural radiation we are subjected to is from radon, 10% from cosmic radiation and about 12% from medical sources such as X rays, and only 0.1% from discharges from nuclear plants.

I have always found the units used to measure the radioactivity we receive very confusing, not least because the units always seem to be changing. I will try to explain.

The *rate* at which the radioactivity of an element decays has been described in Chapter I as its half life of and may be from fractions of a second to millions of years.

The *amount* of radiation - alpha, beta or gamma - produced by one decay per second is called the "*becquerel*" now superseded by the "*curie*". None of this is really helpful when we are considering the effect of radiation on ourselves or on materials. What we want to know is the *dose* we receive. This is measured in "*grays*" (previously "*rads*") which covers all types of radiation. But alphas are much more dangerous (but more easily shielded) than beta or gamma so we need a measure of the biological effect of the radiation on our bodies. This unit is called the "*sievert*" (previously "*rem*" - 1 sievert = 100 rems). The sievert is far too large a unit to use - a dose of one sievert is unlikely to occur except in a major catastrophe. The unit mainly used at present is the millisievert or mSv. - one thousandth of a sievert.

We are all subject to radiation (known as "background" radiation and for a person in the UK this is on average 2 mSv per year (200 millirems). This is made up of 0.3 mSv due to cosmic radiation, 0.35 mSv due to the soil, 0.38 from our own bodies and 0.97 from Radon and Thorium. The amount of man-made radiation is minute - averaging 1.5% of the total.

In order to establish the "safe" level of radiation a very long term study of the effect on the human body is required. Sadly, the longest term subjects are the survivors from the Hiroshima and Nagasaki bombs. To correlate the effects on them with present day exposures requires a knowledge of the strength of the radiation at that time. As no one was likely to have had a "dosimeter" at the time of the explosion, the results have to be extrapolated from other data.

The International Commission on Radiological Protection (ICRP) has proposed that a limit of 1 mSv per year should be the limit of man-made radiation for members of the public and in the UK I understand this is now 0.5 mSv. Remember that we are receiving about 2.0 mSv of background radiation all the time.

What probably concerns us more than anything is the increased risk of developing cancer or leukaemia as a result of exposure to radiation and this is very difficult to answer. There is also a tremendous amount of public anxiety about nuclear power stations. I believe this is partly because people do not understand enough about radioactivity and because it is something they cannot see or feel. The fear of the unknown is very frightening and this is not reduced by some of the propaganda proliferated by the media. The amount of radiation allowed to escape from the operation of nuclear power stations is less than the limit proposed by the ICRP for the population as a whole from all sources of man-made radiation (X rays etc.). This should be considered in the context of carcinogenic (cancer producing) chemicals emitted from coal and oil, and to a lesser extent gas fired power stations. Remembering also that they inevitably emit large quantities of carbon dioxide, contributing to the "greenhouse" effect and sulphur and nitrous oxide, contributing to "acid rain" - (Chapter VII). The larger risk from nuclear power stations is not from their operation but from the processing of the spent fuel and containers used in the reaction and the containment vessels. This will be discussed later under the section on Dismantling. There have of course been incidents, and perhaps that at Windscale was the first to cause public concern.

The Windscale Fire

On the seventh of October 1951 the No. 1 plutonium pile at Windscale was shut down for a routine change of the fuel elements. A little later the pile was started up at low power in order to raise the graphite temperature to release the "Wigner" energy, described previously. This was a perfectly normal procedure. Next morning the control rods were inserted to shut the pile down and later it was started up again at low energy to continue the "Wigner" release. This time something appeared to be wrong as the rate at which the temperature was rising was much too high although the temperature of the graphite was well within its safe limits. The control rods were run in once again to slow down the pile, hoping that the temperature would stabilise. It didn't. The air flow through the pile was therefore increased to cool it but the temperature continued to rise. Next morning as the temperature had still not stabilised and was still rising, the air flow was increased again and this triggered the radioactivity alarm at the top of the chimney. This was not considered significant at the time as a sudden increase in the air flow did frequently sound the alarm, until the flow became stabilised.

However as the temperature and activity levels continued to rise in spite of increasing the cooling air flow and as later high levels of radioactivity were recorded on the roof of the meteorological section, it was evident that radioactive particles were being carried into the atmosphere. This was correctly diagnosed as the failure of the cladding of a fuel element. It was impossible to establish which fuel element had failed as the scanner which would have given this information had jammed!

Eventually the charge channel with the highest temperature was identified. The deputy General Manager, at very considerable personal risk, removed the charge plug and looked inside the channel where the fuel element was glowing red hot. It was impossible to push the element out as the heat had distorted the channel, so the elements in adjoining channels were removed thus isolating the fire - but the temperature continued to rise! Efforts to cool it down with carbon dioxide failed. Water was a better cooling medium to use but in a reactor steam pressure would build up and in the high radiation field it might dissociate into oxygen and hydrogen with explosive results. However as all else had failed, water was turned on in large volumes and eventually the fire was brought under control.

Although the fire was out a new factor was discovered which was a greater potential danger than the fire itself. Samples which had been taken by personnel in monitoring vans around the site, had found that the concentration of radioactive iodine (I 131) in milk samples from Seascale were dangerously high.

Normal non-radioactive iodine (I 127) is needed in our human diet to prevent the enlargement of the thyroid gland, a disease known as goitre. As both I 127 and I 131 are identical chemically, both will go into the thyroid and if the proportion of I 131 is high it can cause radioactive damage leading to cancer of the internal organs. Fortunately the thyroid can only accept a limited quantity of iodine, the excess presumably being excreted. If the concentration of I 131 in relation to I 127 is comparatively low little damage will occur. The cow however concentrates iodine in her milk, so that a cow grazing on contaminated pasture has a higher proportion of iodine in her milk than is in the grass itself.

Consequently whilst it may be dangerous to drink the milk from that cow there would be little danger from eating vegetables (or the grass!) from that area. The level of radioactivity in the milk was considered to be safe for adults to drink but not for children. The difficulty for the scientists was that no recommended tolerance for I 131 in milk had been established at that time. Of course zero concentration would be the safe limit but in an emergency such as this some limit had to be set and it was decided that 0.1 micro curies per litre (the curie was the measure of radioactivity used at that time) would be safe. The samples from Seascale contained 0.8 microcuries per litre. Milk was therefore banned in an area 36 square kilometres surrounding the site. Two days later this was extended to 500 square kilometres - as far as the North Wales Coast.

Not surprisingly the ban on milk caused more public concern than the fire itself, and this was made worse by the secrecy surrounding it. One can appreciate the difficulties for the scientists dealing with this type of emergency for the first time and the taking of samples over such a wide area must have caused many problems. As a result Sodium Iodate tablets are now being made available to the public if such an emergency occurs again. These tablets, containing non-radioactive iodine (I 127) supply the thyroid with sufficient non-radioactive iodine to prevent it absorbing the radioactive isotope (I 131) if it is present.

However one can only hope that the public will be better and more swiftly informed of the potential dangers in any future emergency.

Chernobyl

The reactors at Chernobyl were water cooled with graphite as the moderator. They had instruments for monitoring the reaction and controls to shut it down if it became dangerous. Regrettably these could be turned off or overridden by the operatives. In April 1986, for reasons which are not clear, an experiment was initiated to ascertain what was the maximum power the reactor could produce. To do this the instruments and controls were turned off!

As a result when the reaction got out of hand it could not be controlled and the heat melted the fuel rods. An enormous explosion then occurred, blowing radioactive debris into the atmosphere and contaminating the countryside for miles around. The neutrons escaping and colliding with surrounding atoms, produced isotopes such as strontium 90 and caesium 137 with half lives of many years. Sand and clay was dropped on the reactor from helicopters, nitrogen was pumped in and the whole complex was encased in large masses of concrete.

Perhaps the worst aspect of the disaster was that the lessons learned at Windscale had not been taken note of. The conflicting and delayed evidence of the seriousness of the disaster and the secrecy surrounding it has made the general public extremely suspicious of any "official" advice on such happenings and has probably slowed down the development of nuclear power in this country.

One of the far reaching consequences of these incidents which causes us the most foreboding, is the deposition of radioactive substances on the surface of the Earth from the atmosphere which is known as radioactive "fall-out". There are three types of fall-out from the explosion of nuclear weapons or from nuclear explosions such as occurred at Chernobyl: "local fall-out" as a result of which large particles from the explosion are deposited at a range of approximately 100 miles in the first few hours; "tropospheric fall-out" from which fine particles are deposited around the globe at the approximate latitude of the explosion and within a week or so; finally, "stratospheric fall-out" being the ultimate world-wide deposition, over a period of years, of the particles which were carried by the explosion into the stratosphere. If the reader is not familiar with these definitions of the atmosphere they will be described more fully in Chapter VII on the Environment.

As a matter of interest, the explosion of the volcano Krakatoa in 1883 spread dust into the stratosphere which caused spectacular sunsets for several years. Chernobyl was 100 times worse than Windscale (1957)and 1000 times worse than Three Mile Island (1977) so it was an extremely serious accident which should never have been allowed to happen.

Nuclear Waste and Dismantling

Nuclear power generation must, at first sight, appear to be the answer to our ever increasing demand for power. It has an almost inexhaustible supply of fuel and does not emit polluting gases to the atmosphere. However it does have its disadvantages.

At present in reactors, as we have seen, fuel rods have to be renewed and in some cases plutonium is produced in greater quantities than is consumed. These wastes have to be disposed of and may have active lives, emitting radiation, for hundreds of years or more. They cannot be decontaminated as is the case of those war gases such as mustard gas which can be destroyed by chemical means or by burning. There is no way of stopping a radioactive material from emitting radiation. It simply has to die away in a time depending on its "half life". Fortunately the volume and weight of the waste is very small compared to the large amount of waste from coal fired power stations, for example.

Radioactive waste has to be stored in such a manner that the radiation it emits can be prevented from reaching the environment. This can be achieved by shielding with large volumes of concrete and storage for hundreds of years. The problem here is to ensure that the storage remains intact under any eventuality such as movements of the earth's crust, earthquakes, or even atomic explosions.

The wastes from nuclear power stations are classified as "high level", "intermediate level" and "low level".

The high level waste which arises from the reprocessing of the nuclear fuel is the most radioactive. In the case of the Magnox reactors the spent fuel rods are removed from the magnesium alloy cases and dissolved in nitric acid. The liquid, after reducing its volume by concentration, is stored in stainless steel tanks encased in thick concrete. Because the liquid is still radioactive it produces heat and has to be cooled by water flowing through cooling coils. Although the volume is low - contained in something like 14 tanks at present - the activity is very high and should the contents escape it would constitute a very serious hazard. In order to reduce its volume still further a vitrification plant has been constructed which will incorporate this waste into a solid glass like material which can be stored indefinitely. Glass was chosen because many glass-like semi-precious stones which were formed many millions of years ago, are still intact today.

The reprocessing is carried out at Sellafield in Cumbria where the main source of potential contamination is the liquid effluent passing to the sea. This can cause radioactivity in fish and shellfish which if these are eaten will pass to the human body. The amount of radiation from this source is half the limit proposed by the ICRP and is being continually reduced by improvements to the plant.

One of the great concerns to the workers at Sellafield was that there had been four cases of childhood leukaemia in the area and could this be linked to the radioactivity from the plant? A group chaired by Sir Douglas Black concluded that from the *information available* the radiation from Sellafield, since it commenced operating in 1952, would not be sufficient to cause these cases. However even one case of leukaemia is too many if it can be avoided and research is continuing.

Because the magnesium alloy cans deteriorate rapidly when stored in water after discharge from the reactor, reprocessing of the high level waste from the Magnox reactors is essential. AGR and PWR fuel is in the form of hard ceramic pellets in stainless steel or zirconium cladding and can be kept indefinitely without deterioration.

Originally the need for reprocessing was based on the requirement for recovering uranium as a fuel and plutonium for fuelling fast breeder reactors and for military use. At the present time neither of these criteria is valid and there is controversy as to whether storage should take the place of reprocessing unless or until the latter should become necessary.

The transportation of plutonium in its containers has also become a matter of some concern. It should be remembered that plutonium itself is only mildly radioactive - in fact scientists assembling the first experimental bomb handled the plutonium slugs which they reported as feeling "slightly warm" like a rabbit!. This does not mean that its long term effects is anything but serious. The late Dame Janet Vaughan was the first to recognise the pathological effect of plutonium on the skeleton, writing several books on the subject as well as providing evidence for the ICRP. The intermediate level waste arises from various sources such as the graphite and the cases which have been removed from the fuel assemblies. It does not have as high a radioactive content. It can be cast into solid concrete blocks and stored in special facilities above or underground.

The low level waste consists of a variety of materials which have been in contact with radioactivity, such as clothing, plastic containers, scrap metal, etc. - not only from the nuclear industry but also from medical and industrial sources. Their radioactivity is extremely low but their volume is great. It has been suggested that to be consistent, such items as a bag of Brazil nuts or a bag of fertiliser which emit radiation naturally as much if not more than that permitted by the regulations, should it considered as low level waste!! The low level waste can be compressed and buried in concrete lined vaults under a minimum of two metres of suitable soil on top of a waterproof membrane.

There are also the gaseous and liquid effluents to be considered. Gases are scrubbed to remove any radioactive or chemical material before discharge to atmosphere. The aqueous wastes are treated to remove any radioactive material in solution by a process (flocculation) which brings them out of solution so that they can be removed by filtration. The resulting water which is carefully monitored for any residual radioactivity is safe for disposal via a normal effluent system to the sea. The material removed in the filters is treated as intermediate waste.

Dismantling

When a nuclear power station has reached the end of its useful life, for whatever reason, the question arises as to what to do with it. If it were an ordinary power station or a chemical plant we should just call in the demolition experts and scrap merchants and the whole complex would be razed to the ground. The nuclear power station or plant is very different.

In the case of the Magnox stations the fuel rods will still be in position and will have to be removed for reprocessing or storage. The facilities for this have only been based on so may fuel changes at a time. To process them all may take several years. Then everything which has been in contact with radiation, especially neutron radiation - the steel, the concrete etc. - will be radioactive to a degree from, for example, isotopes such as cobalt 60 in the steel and carbon 14 in the graphite. This would require some degree of protection for the workers and in some cases remote control by robots, causing enormous expenditure of money and of time running into very many years.

An alternative is that the plants should be sealed and left for 130 years!! By that time the radioactivity will have decayed to a safe level. Perhaps before then someone will have found a better way of decontamination!!

Recently it has been reported that the Japanese are already experimenting with a new method of decontamination. This involves subjecting the "high level" wastes to an intense beam of neutrons from a fast reactor such as the FBR at Dounray, or to a beam of high energy protons. This transmutes the elements with very long half lives, which are potentially the most dangerous, into ones with much shorter lives or little or no radioactivity. This means they can be stored with greater safety for a shorter time. Although started in this country, lack of funds means that the research on this project will be carried out abroad.

Fig. 2   Waves on Water

Fig. 14 (a)
Apuan Alps

Fig. 14 (b)
Carrara Marble

Fig. 33 Power Station Cooling Towers

Fig. 35
CERN

Fig.34

Lorenz Atractor

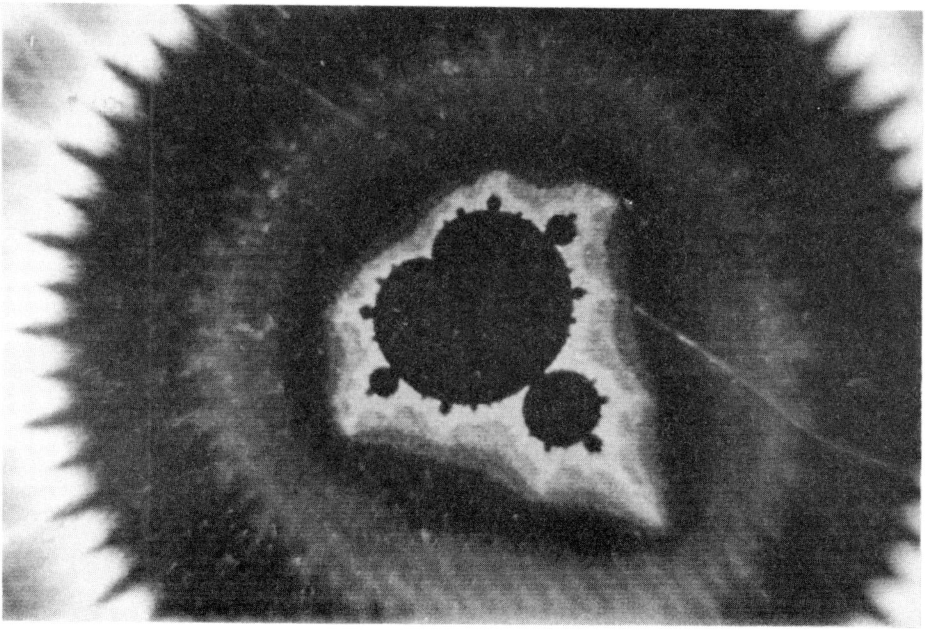

Mandelbrot Set

CHAPTER VI

"WIRELESS", TELEVISION, LASERS AND SUPERCONDUCTIVITY

Wireless

In the 1920s it was called Wireless, presumably because communication by telephone or telegraph had required wires between the sending and receiving stations and this new form of communication no longer required telephone or telegraph wires. The early receiving sets in the home were very simple - mine consisted of a "Cats-Whisker" and a "crystal" and was called, not surprisingly, a crystal set - but more of that later.

Radio waves, as I shall now call them, are electromagnetic waves which I said in Chapter I that I would try to explain.

Fig. 25 Electromagnetic Wave

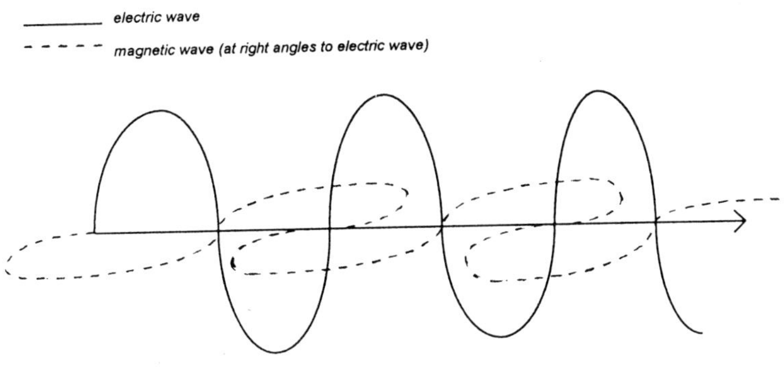

When an electric charge, such as an electron, is accelerated it produces both an electric and a magnetic field which is propagated as a wave motion through space. The electric and magnetic fields are at right angles to each other and also at right angles to the direction of propagation. The fact that they are at right angles to each other is important in future discussions (Fig. 25). The theoretical possibility of these waves was first demonstrated mathematically by James Clerk Maxwell in 1864 but it was Professor Heinrich Hertz at Karlsruhe Polytechnic in the latter part of the nineteenth century who first produced what we now know as radio waves. His simple apparatus which can be considered as the first radio transmitter and receiver consisted of two metal plates each with a knob which when placed close together formed a spark gap. You remember the sparking plug in your car engine. A high voltage source of electricity connected to the plates caused sparks to jump across the gap. While this was taking place, a loop of wire with the ends just separated, was held a short distance away from the plates. Every time a spark occurred between the knobs on the plates a spark appeared at the ends of the wire although there was nothing in between (Fig. 26).

Fig. 26 Hertz's Radio Transmitter

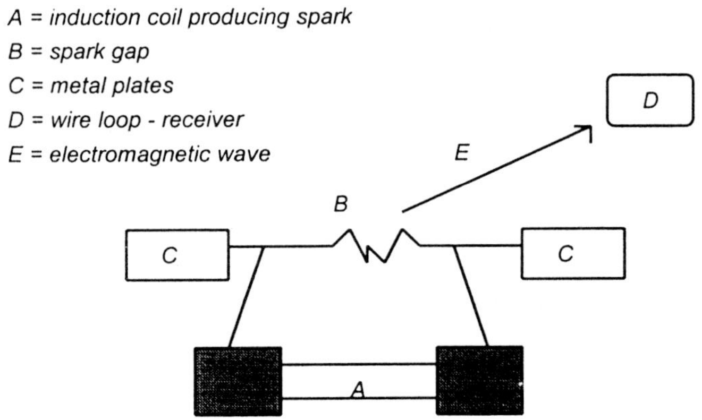

What was happening was not that the spark was producing the electromagnetic waves, but every time the voltage between the plates rose sufficiently for a spark to pass, the plates which can be considered as what we call a condenser, discharged, the spark passed and the other plate became charged in the opposite direction. The process repeated in fractions of a second caused an oscillating discharge which produced the electromagnetic wave.

You can compare this with a pendulum which set swinging will, pass continuously through its lowest point before coming to rest. Or, if you tilt a basin of water and then let it down suddenly, it will surge to and fro until coming to rest. Unless you tip it too far in which case you will get wet! Hertz found that his receiving loop had to be of a certain size and in a certain position to produce the spark. He was in fact "tuning" it to receive the waves in the same way that a piano wire will respond to the sound waves given off by another wire only if it is of the same length and tension as the first wire. He also found that these waves passed through wood and stone and could be reflected in the same way as light but using sheets of metal in place of mirrors.

Electromagnetic waves are propagated through space at the speed of light - 186,000 miles per *second* and cover the whole spectrum of radiation from radio to infra-red, visible light, ultra-violet, X rays, gamma rays and finally cosmic radiation (see Table 1 in Chapter I).

Hertz's experiments produced waves which could only be detected at distances of a few inches which was of no practical use for communication. However, Guglielmo Marconi found that by burying one of Hertz's plates in the ground and using a long elevated wire for the other he could transmit and receive over a distance of 10 miles. His transmitter also used a spark and the electric signal which it produced could be detected by the receiver. In 1897 a ship was fitted with the apparatus and communicated with a shore station 18 miles away. Two years later the distance was increased to 100 miles and in 1901 signals were transmitted across the Atlantic. Such was the speed of development.

The frequency of these early radio waves was of the order of 20,000 hertz (cycles per second) and a wavelength of several miles. The reason for using a spark was that at this time it was the only means of producing electrical oscillations of this magnitude. You might think, why not use the ordinary alternating current generators used in the electric supply industry? Nowadays the supply to your house is alternating current but before the universal supply of electricity many country houses had their own supply which was direct current. However, the generators operate at only 50 or 60 hertz and to make an alternator to produce even 20,000 hertz was at that time a difficult and expensive technical problem.

The invention of the thermionic valve by Dr J. A. Fleming which I will now describe produced the solution. Most of the electric light bulbs you buy now are "pearl" or opalescent so that you cannot see inside them. However if you buy a clear one you will see inside that there is a very fine metal wire or filament which now is generally made of tungsten because of its high melting point so that it will not melt at its operating temperature. This filament is heated by an electric current so that it becomes incandescent and emits light. The bulb is evacuated and all air removed as otherwise the filament would burn up and be destroyed.

You will have noticed that electric light bulbs eventually blacken and burn out. This phenomenon was first investigated by Thomas Edison who had produced the first satisfactory light bulb with a filament made of carbon or graphite. As a matter of interest he had also designed the first commercial lighting system in New York and on the 1st of September 1882 he switched it on for the first time.

One of his ideas to overcome the problem of blackening was to place a second electrode into the bulb, the filament being the other.

He noticed that if he put a battery between these two electrodes a current would flow but only in one direction. At the time he could see no application for this but Dr J. A. Fleming, having seen the demonstration, took the matter further.

He placed a small hollow metal cylinder round the filament, but not touching it now, known as the plate or anode and connected it to a terminal outside the bulb. He found on switching on the current and connecting a battery between the plate and filament that a current flowed so long as the plate was connected to the positive terminal of the battery but no current flowed if the polarity was reversed.

This effect is caused by the electrons, which are ejected from the filament on account of its temperature, being collected on the plate, allowing current to flow, the movement of the electrons being the current. If the plate is connected to an alternating current it allows the current to flow in one direction but not the other, in much the same way as the valve on a motor car tyre allows the air to flow into the tyre but not out. The device was therefore called a valve. Subsequently Dr Lee de Forest interposed a fine wire mesh between the filament and the plate and found that an electric potential applied to this "grid" could control the amount of current flowing from filament to plate (Fig. 27).

This then allowed the valve to be used as an amplifier because a small potential or current applied to the grid allowed a much larger current to flow through the plate. Further, by adding various external components, the valve circuit can be made to oscillate producing an alternating current. So here we have a means of producing the high frequency, high power oscillations required to produce the radio waves.

Fig. 27 *Early Thermionic Valve*

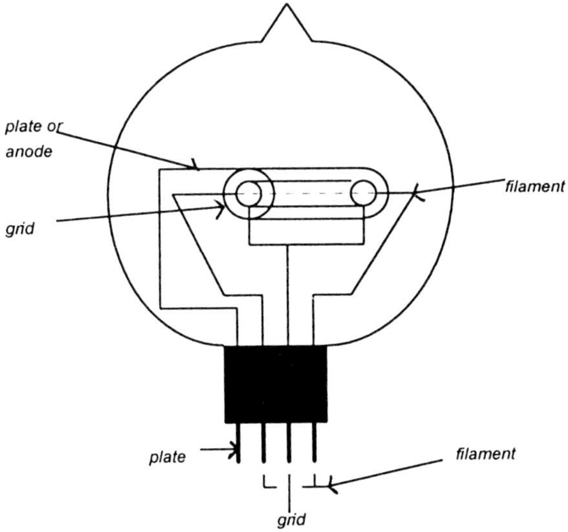

The original transmissions by Marconi consisted of switching the transmitter on or off for long or short periods which by means of the Morse code was able to convey information to the receiver. In order to transmit speech - and later television - this is not sufficient. The signal must be infinitely variable to follow the changes in current produced, for example by a microphone which transforms speech into different levels of electric current. The radio wave is a continuous one of a definite frequency depending on the transmitter. If this were to be received, nothing would be heard as the frequency is far above what the human ear can respond to. Even if its strength were varied by the microphone, nothing would still be heard as the wave varies equally first positive and then negative and the result would cancel each other out. If, however, we cut out either half of the wave then the resultant signal will be significant and the variations applied at the transmitter can be heard at the receiver.

To return to my crystal set. It comprised an aerial wire about 100 feet long, strung up between but insulated from, two fruit trees in the garden and an earth plate consisting of a metal rod driven into the ground. The set consisted of a coil of wire which could be adjusted to receive the required wavelength of the radio signal and was connected to the aerial and the earth. Also connected in the circuit was a pair of headphones and a crystal of galena (lead sulphide) with a fine gold or silver wire - the Cat's-Whisker - which could be moved over the surface of the crystal. The galena which had been formed by crystallisation, had a number of minute points of contact on its surface which would conduct an electric current in one direction only. It was the purpose of the Cat's-Whisker to locate one of these points and it was a great achievement when scratching over the surface a signal was heard.

In those days the Manchester broadcasting station had the call sign 2ZY and for London 2LO. I was about 25 miles from Manchester so frequently received 2ZY but on one memorable occasion I heard 2LO. This was not the only thing I heard. My aerial was rather close to the public telephone wires along the road and at times my receiver picked up some local telephone conversations which, although interesting were, I suppose, an early form of illegal "phone tapping"!

I next advanced to a valve receiver. The original valve which I still have was a "Bright Emitter" (see Fig. 27), so called because the filament glowed brightly, but required a large storage battery (car battery) to keep it going and a dry battery of some 100 volts for the plate. Later filaments were produced which operated at much lower temperatures and currents, these valves were known as "Dull Emitters". Valves not only rectified (i.e. detected) but also amplified the signal which could then be fed into a loudspeaker. How far we have advanced since then!

Directional transmitters now beam the radio waves across the world, some hug the surface and some are reflected from the upper atmosphere and some set out into space where they are still travelling at speeds of 186,000 miles every second to who knows where.

Radio waves normally travel in straight linen from where they are generated. It would not have been possible to receive them at places out of sight of the transmitter were it not for the fact that they can be reflected or bent by certain layers of our atmosphere.

What is called the Kennelly Heaviside layer (after its discoverers) is a part of our atmosphere, approximately 55 to 85 miles above the surface of the Earth and containing free electrons produced by the impact of ultraviolet radiation from the Sun on the atoms of our atmosphere and also from the trails of ionised gas caused by meteors entering the Earth's atmosphere. It varies from day to night and causes the "fading" of radio signals which was a problem in the early days.

Now that we have "Satellites" many miles above the Earth, radio waves are "beamed" to them and reflected or retransmitted to the other side of the Earth and are no longer subject to fading.

Modulation which means the impression of the required signal on the continuous or carrier radio wave was as already described carried out by varying the amplitude or size of the carrier wave. This method has now been mainly superseded by frequency modulation (FM) which by varying the frequency of the carrier has many advantages including increased quality of the sound and absence of fading but at the expense of more complex equipment. However the most significant advance was the invention of the transistor.

The Transistor

The invention of the transistor is attributed to a team led by Professor John Bardeen, William Shockley and Walter Brattain in 1948 during an intensive study of semi-conductors.

In order to understand the transistor we must first consider the diode and semi-conductors in general.

The diode is a device which allows electric current to flow through it in one direction only - i.e. it is a rectifier like the thermionic valve already described and the galena crystal in our "crystal set" but it does not require a large current to heat a filament nor a high voltage for the plate as does the valve.

You may not have seen a diode but there is one in your car where it is used to rectify the current from the alternator. Most cars now have alternators in place of dynamos. The dynamos produced a direct current which was necessary to charge the car battery and so did not need rectifiers. Physically the diode is a very small component which you may never have noticed! An alternator produces an alternating current which would be no use for charging a car battery so the rectifying diode is required to change the current into a direct one.

I have described in Chapter I how electricity flows in a conductor such as copper which has sufficient free electrons to carry the current. Where there are no free electrons the material is called a non-conductor or insulator such as rubber or glass. In the case of substances such as silicon (made from silica sand) or germanium which are known as semi-conductors there are few free electrons at room temperatures and they do not conduct appreciably. We say their resistance is high.

You will remember when we described the atom it had electrons rotating in orbits round the nucleus and when there were many electrons they fitted themselves into what we can think of as spherical shells. The outer shell of a silicon atom contains four electrons, the normal number in the outer shell of an atom would be eight. In the silicon crystal which is grown from the elemental silicon by a very special technology, each silicon atom shares its four outer electrons with four from adjacent atoms and is stable and does not conduct electricity appreciably (Fig. 28 (i)).

Fig. 28 (i) Silicon Crystal

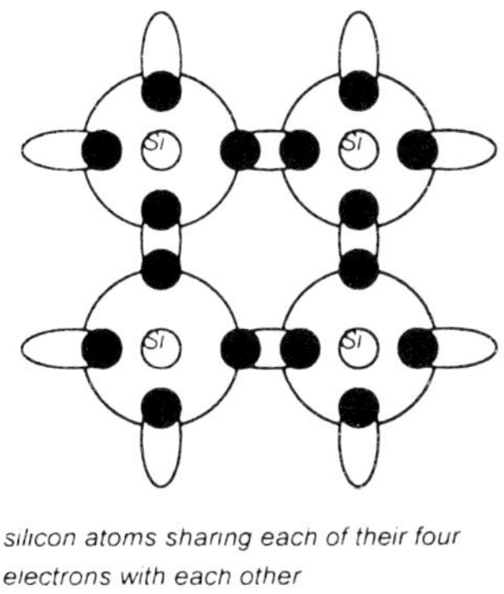

Fig. 28 (ii) Silicon Crysal doped with Phosphorous

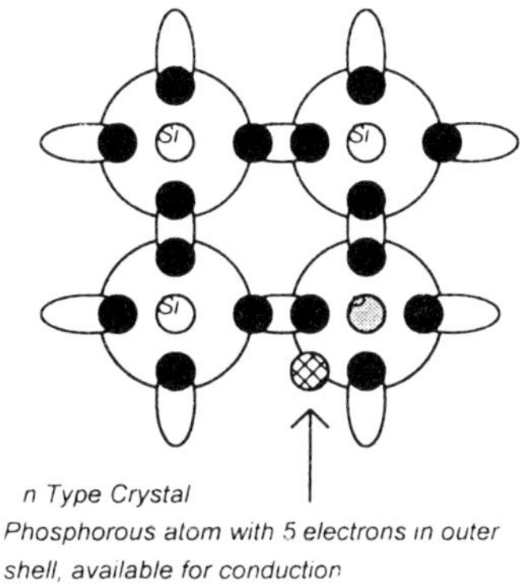

Fig. 28 (iii) Silicon Crystal doped with Boron

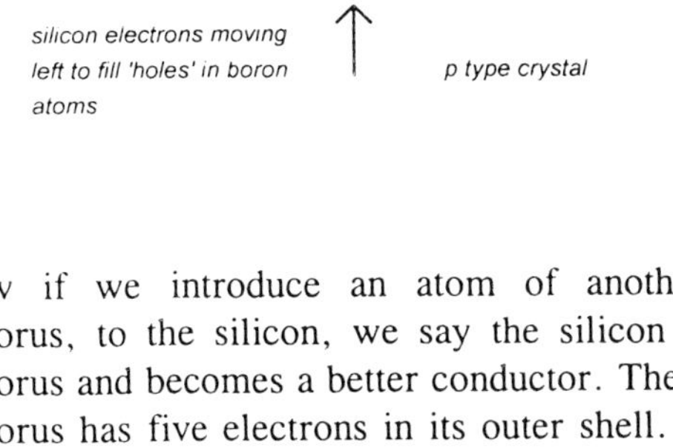

Now if we introduce an atom of another element, such as phosphorus, to the silicon, we say the silicon has been doped with phosphorus and becomes a better conductor. The reason for this is that phosphorus has five electrons in its outer shell. Sharing four with the silicon leaves one free to conduct and we have made what is known as an *n-type* semi-conductor (Fig. 28 (ii)).

If instead we introduce an element such as boron, which has only three electrons in its outer shell there is going to be one missing from the eight required, leaving what is known as a hole. This hole can be filled with an electron from another silicon atom but this creates another hole and so on throughout the crystal. The silicon will now conduct in the opposite direction and is said to be a *p-type* semi-conductor (Fig. 28 (iii)).

Now to try to make it as simple as possible. A diode is a piece of silicon (or germanium) which has first been doped with phosphorus to make it n-type. Then one surface is exposed to boron vapour in a furnace. The boron atoms diffuse into the surface of the silicon and part of it becomes p-type. There is a very thin layer between the two containing both p- and n-types which acts like a little battery (0.6 volts in the case of silicon and 0.2 volts for germanium) (Fig. 29).

Fig. 29 pn Junction Diode

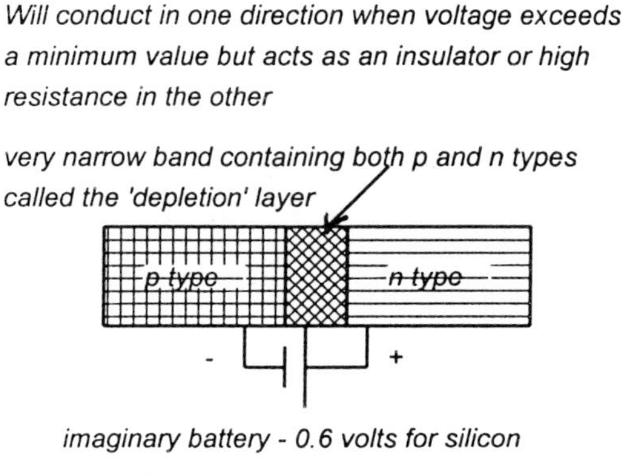

When we apply a voltage to the device, as we shall now call it, current will flow in one direction but not in the other, as soon as the voltage exceeds 0.6 volts in the case of silicon. So we now have a means of rectifying an alternating current by applying it to a diode.

You may well ask why we talk about "n" and "p" types. I have been told that the origin of this is that when an impurity adds electrons to a semi-conductor it is classed as a "do(n)or" and hence is called "n" type. Similarly when it has a surplus of (p)ositive charges it is classed as an "acce(p)tor" of electrons and hence a "p" type. What a convoluted way to arrive at a nomenclature!

But what about the transistor? We start with a slice (or chip) of silicon (which is n-type) and produce a layer of silicon oxide, which is non-conducting, on it. Through a hole in this oxide the silicon is exposed to boron in a furnace and a layer of p-type silicon is produced as in the diode. This is covered with another layer of oxide and a hole through this exposed to phosphorus producing a layer of n-type material. We now have a sandwich of n-p-n material (known as an n-p-n transistor) which is in effect two diodes connected back to back, with external connections to base, collector, and emitter, as they are called (Fig. 30)

Fig. 30 *n-p-n Transistor*

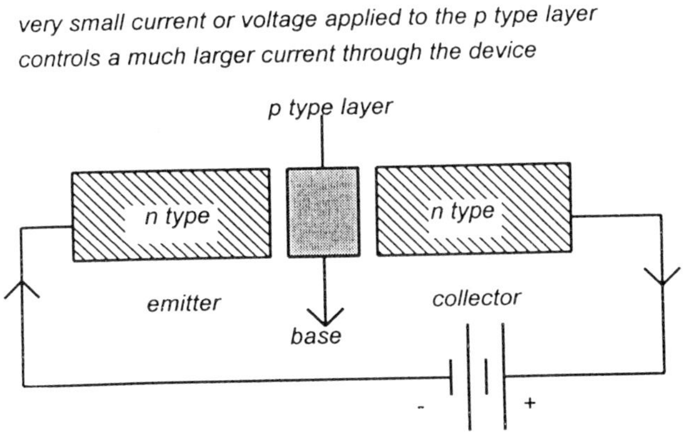

The tremendous value of this device is that by applying a voltage, or more correctly, a small current, to the base connection the current between the collector and the emitter can be controlled - it acts as a variable resistor where the effect of variations in the very tiny base current are transferred to the collector current. Originally it was called a transfer-resistor and hence shortened to transistor. There are literally hundreds of types of transistors from the very small to the large current carrying types and we can truly say that they have transformed our lives. Their small size and low power requirements have meant they have supplanted the thermionic valve in all but a few applications and their development, not only in radio, but also in all walks of life has made much that before was impossible now commonplace.

Before leaving the transistor, I must mention the Integrated Circuit (or chip). With the development of the technique for producing the transistor, it was found that numbers of transistor junctions could be etched and interconnected on one slice of silicon so that hundreds and thousands of transistors could be produced on a single chip not much larger than the original diode. Not only transistors but other components such as resistors and condensers could also be produced on the chip.

Now we have devices, measuring not more than two or three centimetres, with thousands (and now millions) of components including memory cells which form the basis of our computers, calculators etc. and to this at the moment there seems no end.

Television

"Wireless" was a great breakthrough at the beginning of the 20th century. People could speak to each other across the oceans of the world. Australia was no longer a continent at the bottom of the world, as viewed from England, and America was just a flash away across the Atlantic. However, people could not "see" their friends at the other end of a telephone line. So this was something the scientists and inventors in England and America put their minds to.

Various ingenious electro-mechanical devices were considered and constructed and eventually a still picture or a photograph could be transmitted along a telephone wire. This was called "phototelegraphy" - we now call it "fax"!

True television is the instantaneous transmission of a moving scene as viewed by the eye or the lens of a television camera.

The first successful demonstration of television transmission along a cable was carried out by John Logie Baird on January 27th 1920 at the Royal Institution. His equipment, which is now in the Science

Museum, looks like a collection of old bicycle sprockets, cardboard discs and bullseye lenses - very far from the sophisticated TV of today. But it worked.

The basic principle of television was, and still is, to split the image into a very large number of small parts. If you look at a picture in a newspaper under a magnifying glass, you will find it is made up of a number of small dots of differing density of black to white or of differing colours, but when you view it normally with the naked eye it is one complete picture.

If you can then measure the intensity of the light in each spot, convert it to an electric signal, transmit it to the receiver, and then reproduce the spot in exactly the same place and with the same intensity and do the same with all the spots - you have a picture! Then if you want to have a moving picture you have to repeat this process at least 14 times every second. It all sounds very complicated but Baird achieved it without our modern electronic devices.

Television like cinematography depends on an inherent property of your eyes known as persistence of vision which means that so long as an image is repeated faster than about 14 times per second we see it in continuous movement.

In the modern TV system the first problem is to convert the visual image into electrical impulses. The image is focused by lenses similar to those in your camera onto a plate of photosensitive material whose electrical characteristics depend on the intensity of the light falling on it. Behind this is another plate of transparent material but which is also a conductor of electricity. If a potential is applied between these plates a current will flow which will vary with the intensity of the light striking the photosensitive plate - rather like the microphone which transforms sound into an electric current. But this is of no use by itself as it would only give a result which was due to the average intensity of the light over the whole image.

Behind the photoconductive plate there is what we might call a small cathode ray tube (see Chapter I). This projects a beam of sharply focused electrons onto the plate.

The trick is that the beam is made to travel from left to right tracing a series of imaginary lines on the plate. When it reaches the bottom right it moves rapidly to the top left and "scans" the plate again. Each time the beam strikes the plate it varies the plate's electrical property at that particular point in proportion the intensity of the light falling on the other side and so produces a signal proportional to the intensity of the light on that spot. This signal can then be transmitted either by wire or radio to the receiver (Fig. 31).

Fig. 31 Television Camera

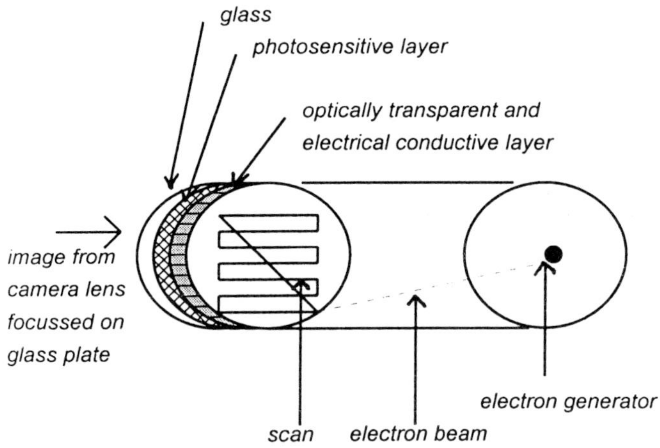

Your TV receiver consists of a tube which again is a cathode ray tube with a heater or filament to produce the electrons and a very high voltage to propel them (25 thousand volts - so don't open the back when the power is on as 25 Kv can be lethal!). The electrons strike the inside of the tube at the front of the set. The inside is coated with a material (a "phosphor") which emits light in proportion to the strength of the electron beam striking it which has been modulated by the signal from the transmitter. (Incidentally, most of the electrons are stopped by the glass of the tube but it is still advisable NOT to sit close up to the screen.)

The beam, as in the case of the television or video camera, scans the picture with 625 lines and scans the whole frame 25 times per second. (This is the European standard although the scan is doubled to 50 per second to reduce flicker - in the USA the standard is 525 lines and 30 frames per second.)

Now this may seem fairly simple but the catch is that the beam in the receiver must be exactly synchronised with that in the camera or you will not get a picture. This is accomplished by special "sync" signals transmitted along with the picture signals.

You now have some idea of the speeds involved by the electron beams in producing a picture. The sophisticated electronic circuits now required for our modern TV would not be possible without the transistor and the integrated circuits (ICs, already described). Baird had none of these and the early sets used thermionic valves which werc bulky, fragile and hot.

So far what I have described would be a black and white TV. Colour TV is more complicated. It depends on the principle that any colour or hue can be obtained by mixing the three primary colours, red, green and blue, in differing proportions.

In the camera the picture is split into these three colours by an ingenious system of special mirrors which only reflect the coloured light required. The scan is three times as fast and three separate signals are produced. In the receiver there are three electron guns each of which takes one of the primary colour signals. The inside of the television tube is coated with a series of vertical strips of selective phosphors which give off the primary colour when struck by the electron beam. These strips are very close together so that the eye cannot differentiate between them.

The three guns have to be most carefully focused so that they only strike their own strips.

All this means that the electronics are extraordinarily complicated and whenever I see a colour TV set I am amazed that it works as well as it does!

Lasers

The word laser stands for "Light Amplification by Stimulated Emission of Radiation". If someone asks you - "Do you know anything about lasers?" you might reply - "Oh. You mean light amplification of stimulated emission of radiation" which would be a case of "one upmanship", if you can remember the tongue twister!

Lasers had been thought of for many years. Professor Einstein, as far back as 1917, had suggested that atoms could be stimulated to emit light but it was not until 1960 that Theodore Maiman in California constructed the first ruby laser.

In the last section on transistors I described how the electrons were in shells around the nucleus of the atom. If an electron is moved from a lower (nearer the nucleus) orbit to a higher one energy is absorbed. On the other hand if an electron jumps from a higher to a lower orbit energy is emitted.

In a laser the electrons are "pumped" to a higher level by the application of energy in the form of light or an electrical discharge. This causes an abnormal number of electrons to be in the higher orbits. When these electrons jump to a lower orbit they not only emit energy themselves but "stimulate" other electrons to emit energy as well. A sort of avalanche effect.

Fig. 32 Maiman's Ruby Laser

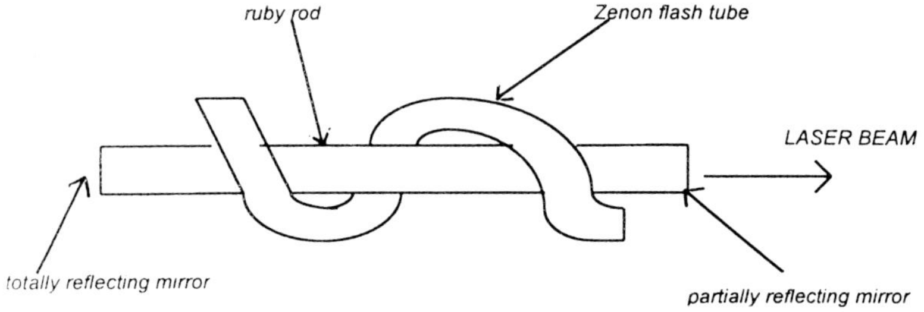

Maiman's first ruby laser was comparatively simple. He put a small ruby rod with flat silvered ends inside a spiral shaped flash tube similar to those used in flash photography but very much more powerful (Fig. 32).

Ruby is a form of aluminium oxide contaminated with compounds of the metal chromium. When the flash tube is fired the chromium atoms absorb energy from the intense flash of light and the electrons are pumped to the higher level. On falling back they release their energy as red light and at the same time stimulating more atoms to release more electrons and so on.

The light is reflected to and fro by the silvered mirrors at the ends of the rod thereby increasing the number of collisions with the chromium atoms and amplifying the effect. The mirror at one end of the rod is only partially reflecting. Rather like the "one-way" mirrors used in certain seedy establishments which, coincidentally, also use "red lights"! This allows the laser light in the form of a beam of red light to emerge.

There are some properties of laser light which are important and which we should consider.

It is *coherent* which means that the undulations of the waves are in step. As an example, a body of soldiers or policemen marching in step is much more effective - so we hope! - than a rabble of demonstrators moving in all directions.

The resultant beam is highly concentrated. It does not diverge as a light beam would, no matter how well focused. Those of you who remember the searchlights in the last war will have noticed that although they appeared to be sharp beams of light in the night sky, they were quite wide. Nowadays if you see a display of laser light in the sky, that light appears as narrow pencils of uniform size until they disappear in the distance.

A magnifying glass can focus the sun's rays so that they are concentrated at a point where a piece of paper will burn, but this only occurs at one point. A laser beam, if strong enough, would burn the paper at any point along its path.

To give you an idea of this. A 1 watt laser with a beam say 1 mm in diameter would at 10 metres only spread to perhaps 5 mm but it would still have the power of 1 watt. A 100 watt electric light bulb, if not concentrated, would at 10 metres only have a power of 1 millionth of a watt over an area of 5 mm.

After the ruby laser very many different types were developed. Perhaps the most common was the helium-neon laser where a mixture of helium and neon gases at very low pressure are subjected to an electrical discharge. This also emitted a red light and is the laser used in reading the bar codes which you have no doubt seen in supermarkets. That bright red light across which the shop assistant passes your purchase is the laser beam and hopefully the correct price appears on the till!

Various devices can be introduced into the laser cavity (between the mirrors) which allow the beam to be "tuned" to very precise wavelengths. Similarly the amount of "pumping" can. be built up and then released in "pulses" of energy of very short duration but exceedingly high power.

In the section on the transistor I have described the diode as a rectifier but it also has another property. Under certain conditions it will emit light and is known as a light emitting diode or LED. In this case the light caused by the electrons is from spontaneous emission.

However, if the diode is constructed in a certain way it can act as a laser and stimulated light emission takes place. These semi-conductor lasers are of tremendous importance on account of their small size and low power consumption. Their cavities can be tuned and pulsed just as in the other lasers.

There are many other exotic types of laser. Neodymium being one. You have probably never heard of neodymium, which is a rare metal (atomic number 60) present in the rare mineral Monazite. This is mixed with an yttrium-aluminium compound. Yttrium being another rare metal (atomic number 39). This laser is much used in cutting operations about to be described.

The uses to which lasers can be put are legion and perhaps are the most interesting aspect for the general reader.

Because of the ultra fine beam and the highly concentrated power in it, lasers are used to cut and drill all manner of substances with a precision impossible by other means. Drilling exceedingly fine holes in diamond is one example. Working titanium metal, used in the aircraft industry was difficult with conventional tools but can be overcome easily by using lasers.

The fine beam of the helium-neon laser is used by surveyors for accurate measurements over long distances and for example, checking the accuracy of tunnels. Certainly used by the engineers of the Channel Tunnel which ensured that the French and English sides joined accurately in the middle.

Lasers are used for communicating with satellites. When reflected back to earth they can allow precise measurements of the movement of those geological "plates" below the surface which lead to earthquakes.

They are also used in the production of Ordnance Survey Maps, making the "trig." points around the country obsolete. A subject which has caused some controversy by some conservationists.

An interesting use of lasers was in the communication with a submerged submarine from a satellite. The laser is tuned to the blue-green wavelength which is least absorbed by seawater hence the blue-green colour of the sea and is pumped to a very high energy in the satellite. The beam can be modulated with data and this is transmitted to the submarine.

Lasers are also used in the remote monitoring of atmospheric pollution in the upper atmosphere.

Surgery has in many cases been revolutionised by the use of lasers in sensitive areas such as the eye, the larynx, and the female reproductive tract.

Perhaps their greatest use is in the handling of information ("IT"). An infra-red beam from a tiny semiconductor laser 1 mm in diameter reads the codes on a Compact Disc music player. The beam can be focused on a spot one thousandth of a millimetre across so that a 12 centimetre disc can store 75 minutes of music, or in the case of a computer it can store and retrieve over a million words on a similar sized disc.

Optical fibres are now supplanting telephone cables. The principle on which they work is another subject but suffice to say that the signal, be it telephone or television, is converted into packets of light which are then transmitted via these fine glass filaments. How light which normally travels in straight lines is persuaded to go round corners is something which I will not at the moment try to describe. However, because laser light is coherent and of one frequency this makes it more efficient in this application.

Lasers can be made to operate at wavelengths from the infra-red to X rays although it is much harder to generate the shorter wavelengths. Their energy can extend from less than a thousandth of a watt such as is used in the supermarket scanners (which can do you no harm) to enormous powers.

The US armed forces have been interested in lasers since 1970 when with a "Mid Infra-Red Advanced Chemical Laser"(MIRACL) which generates more than two million watts for a few seconds, they were working on methods of shooting down enemy missiles - hence "Star Wars".

More recently we have seen them used in the Gulf War for the accurate targeting of missiles.

Probably the most powerful laser is the Nova at the Lawrence Livermore National Laboratory in California. It produces 100 trillion (100,000,000,000,000) watts in a pulse lasting only a billionth (0.000,000,001) of a second. During that brief pulse Nova's power is said to be equivalent to 100,000 power stations and is used for research on nuclear fusion (see Chapter IV).

Lasers can be used in isotope separation. In Chapter III you will remember the enormous effort put into separating the uranium isotope, U235 from the more plentiful U238. Using a copper vapour laser, pumping what is known as a "dye" laser, very accurate wavelengths can be produced which can be tuned to the frequency of a particular uranium atom. As a result U235 can be separated as a vapour at a very high temperature. This may eventually transform the methods of fuelling nuclear reactors and for production of nuclear weapons if they should proliferate.

Although Nova produces exceedingly short pulses, even shorter pulses of laser light can now be made. In 1960 Maiman's ruby laser gave pulses of only a few thousandths of a second. The shortest pulses produced so far are of the order of 8 femtoseconds (8 divided by 10 with 15 zeros after it!). This corresponds to only four cycles of visible light if you can imagine such a short time!

Finally we have been told that the Russians are working towards a huge satellite capable of beaming solar power to Earth. The power would be transmitted via infra-red laser generators. Hopefully about 500 megawatts could be produced - a really "clean" source of power.

Superconductivity

You may have heard about superconductivity in connection with ideas of perpetual motion, so I will try to explain what it is.

A temperature reading tells us how hot or cold a body is. It is basically a measure of how fast the molecules, atoms and electrons are vibrating or moving. The higher the temperature the faster they move and conversely the colder it is the slower they go.

Conventionally we take the boiling point of water as 100 degrees Celsius (it used to be Centigrade or 212 degrees Fahrenheit) and the freezing point of water as 0 deg. C (or 32 deg. F). It is generally measured by the expansion of mercury in a glass tube which you know as the thermometer with which the doctor takes the temperature of your body - and tells you to go to bed! - unless it is normal which is 98.4 deg. F. Now they don't use a thermometer but a specially treated piece of paper which changes colour with temperature.

If we cool a body down to a temperature at which vibration of electrons all but ceases, we reach a temperature of minus 273 deg. C which we call "*Absolute Zero*" and temperatures on this scale are called degrees Kelvin (K) - so water freezes at 273 deg. K and boils at 373 deg. K.

Absolute Zero or 0 deg. K is a temperature below which it is impossible to go as we cannot envisage an electron with a negative velocity. It is, in fact, impossible to reach absolute zero because at a temperature just above it - 0.001 deg. K - the electrons still have a small residual vibration. They are never still!

To give you an idea of these temperatures, the air which we breath can be liquefied at a temperature of around minus 180 deg. C (or 93 K). Our domestic deep freezers on the other hand are only at about minus 18 deg. C.

The most difficult element to liquefy is helium which has to be reduced to minus 268.9 deg. C or only about 4 deg. K above absolute zero. Below this temperature it still remains liquid and cannot be solidified except under great pressure.

Liquid helium has some remarkable properties. It has practically no viscosity nor surface tension (the property which makes a liquid adhere to the sides of a glass vessel for example). If you were to put some liquid helium in a cup, you would find shortly that the helium liquid had crept over the sides of the cup and spilled over, eventually emptying the cup, apparently defying the laws of gravity!

In a conductor of electricity such as a metal wire, the current is carried by the movement of free electrons as we have described in Chapter I. At normal temperatures the movement of the electrons is impeded by the vibrations of the atoms in their path. We call this the resistance of the wire and as a result energy is required to maintain the current and this appears as heat or electromagnetic energy. As the temperature rises, so the number of collisions increase and so does the resistance. Conversely if the temperature falls the resistance decreases.

As we approach 0 deg. K the vibrations of the atoms and electrons all but cease and there is no resistance. A current of electricity, once started, will go on for ever! This is what we call superconductivity. Just imagine how much energy this would save in the power distribution along all those electricity cables and how much more efficient would be the power station?.

Sadly from a practical point of view, except in certain cases, the cost of keeping the conductor at the temperature of liquid helium far outweighs the advantages of nil resistance. There are however some exceptions. In the case of the electromagnets used in the "torus" which we have described under power generation by fusion, it is well worth while cooling in liquid helium to reduce the electrical resistance in the windings and so to reduce the power required and hence the heat developed.

Latest research, which is continuing, has shown that certain materials which are not conductors at normal temperatures do exhibit the property of superconductivity at temperatures higher than 0 deg. K but still far below the freezing point of water. The cost of cooling is then not so important and the possibility of practical applications are exciting - in power generation, transmission and transport as well as a new generation of computers.

One unusual effect of superconductivity, known as the Meissner effect, is that whereas a magnetic field would go straight through a non-magnetic material, in a superconducting medium the magnetic lines of force are distorted so that they avoid going through it. For example, in a circular ring of material the field would not go round the ring but through the centre.

Because this effect allows minute changes in magnetic fields to be measured (one part in a billion of the Earth's magnetic field) it can be used in cardiography, neurophysiology of the brain and the investigation of other bio-physical systems and also in space flight navigation.

There is no doubt that in the future the use of these special materials, which so far are mainly ceramic or clay-like in composition, will become of increasing importance.

CHAPTER VII

THE ENVIRONMENT

The Ozone Hole

The subject of the "Ozone Hole" is an emotive one and one which causes much media coverage and concern to all of us. Although to do with chemistry, it also has a bearing on the action of radiation on the atom as you will see.

Ozone is a different form of the oxygen in the atmosphere which we breath. Normally two atoms of oxygen combine together to form the molecule O_2 but sometimes three atoms will combine to form ozone - O_3. Our atmosphere consists of approximately one fifth part of oxygen and four fifths of nitrogen and a very small proportion of other gases such as argon, neon, xenon, krypton and helium. Although there is only a small proportion of these rare gases they can be separated and are of considerable industrial and scientific use. There is also a certain amount of carbon dioxide (CO_2) which is important in connection with global warming, which we shall discuss later. Ozone can be made by subjecting oxygen to an electric discharge such as a spark. I used to think that the rather acrid smell which was present in the underground railways at one time was caused by the large sparks which occurred on the rails under the coaches. It was in fact produced intentionally as a germicide to prevent the spread of infection amongst the large number of people in a confined space. However it was later considered that it would probably harm more people than germs and was discontinued. There is said to be a slightly higher concentration of ozone in the air by the seaside which no doubt leads to the saying "taking the ozone" by a walk by the sea. The fact that ozone can be produced by the effect of ultraviolet radiation on the oxygen atom is what we are concerned with in what follows.

UV radiation (I will use the abbreviated form UV) which reaches the Earth's surface from the Sun can be harmful but it can also be beneficial. Some of the effects can be very advantageous to the body as they help to produce vitamin "D" in the cells below the surface of our skin; so people who stay too much indoors may suffer from vitamin "D" deficiency. If we are exposed to too much UV the body has an in built defence in that a thicker layer of dead cells is built up which prevents more radiation from entering the body; our skin becomes brown, more so if it is pale already. Further exposure to UV of longer wavelengths can produce a non-malignant form of skin cancer but certain wavelengths can also produce a malignant form - melanoma, although this can be produced without exposure to UV.

If all the UV radiation from the Sun were to reach the Earth's surface, the result would be catastrophic and life would cease to exist. Fortunately when the UV strikes an oxygen molecule its energy is absorbed and an ozone molecule is produced. This happens in our upper atmosphere - the stratosphere - some 7 to 20 miles above the surface. The result is that the amount of UV reaching us is reduced.

However it is a fact that ozone also absorbs UV and becomes an oxygen molecule.

For some time it confused me as to why we were so concerned about ozone removal as the oxygen was absorbing the UV and turning into ozone, and the ozone itself was also absorbing UV and producing oxygen. The answer is that the reactions are selective in the wavelengths they absorb. You will remember from Chapter I and the electromagnetic spectrum, that the UV wavelengths stretched from 50 nm to 400 nm. The reaction of oxygen to ozone takes place mainly with UV below 190 nm and to a certain extent up to 240 nm but *not above* this wavelength. From 230 nm to 290 nm and to some extent up to 300 nm the ozone absorbs the radiation preventing it reaching the Earth. It happens that these wavelengths absorbed by the ozone are the one's which cause the trouble and which is why we are so concerned about the depletion of the ozone layer.

Several natural phenomena can affect the balance between oxygen and ozone. One of these is cosmic radiation which we shall discuss in the next chapter, but the effect is small unless there is a sudden and enormous increase in its activity. It has been suggested that such a thing might have happened some 65 million years ago when a star, exploding in the vicinity of our Solar System, caused a tremendous burst of radiation, destroying the ozone layer and thereby allowing massive amounts of UV to reach the Earth's surface, destroying all life (including the dinosaurs). But life must have started again or we wouldn't be here. There are other natural causes which may remove the ozone layer - volcanic eruptions, thunderstorms, methane from swamps and from the stomachs of animals (including humans) and also methyl chloride which is produced in forest fires and from marine organisms in the sea. None of these natural effects could account entirely for the depletion of the ozone in the stratosphere.

Experimental work over the past two decades has shown that this depletion is increasing. From 1982 the researchers found that the ozone in an area over the Antarctic South Pole was diminishing dramatically which has lead to it being called the "Ozone Hole". The question was - why was this occurring? The simple answer appeared to be chlorine. Chlorine is a very reactive gas and breaks down the ozone to form chlorine oxide and oxygen - so the ozone disappears. The temperatures of the Antarctic stratosphere are very much colder than anywhere else on Earth partly because the atmosphere is more quiescent in the southern hemisphere. Due to the rotation of the Earth we have a vortex (like a whirlpool) over the Pole and in the "eye" of this vortex the air is practically stationary and gets extremely cold in the long South Polar winter. This cause the formation of minute ice particles (Polar Stratospheric Clouds - PSCs). These remove chemicals from the atmosphere which allow free chlorine to form which then reacts with and removes the ozone.

Now where does all this chlorine come from? In the 1920s a series of compounds were discovered, containing chlorine, fluorine, and carbon which were extremely stable, non-toxic, non-flammable, and were ideal for use in refrigerators in place of ammonia gas which was then in use and was flammable and toxic. They were known as "CFCs" (chloro-fluorocarbons). In the 1950s they were used as the propellant in "spray cans" for all sorts of products, and for use in fire extinguishers, foam insulation for cavity walls, solvents, plastics such as hamburger boxes and many other uses. As they were so stable, it did not seem that they could be a threat to the environment. Alas it now appears that, like many other stable compounds, they were decomposed when subjected to very strong UV radiation and consequently released chlorine. How do they get to the stratosphere as they are considerably heavier than air? By the wind and convection (hot air rises). They may take as much as five years to reach the upper layers from the surface of the Earth but once there they may remain for a very long time (a lifetime of 200 years has been quoted) as potential chlorine suppliers.

These results have caused a ban on the manufacture and use of CFCs and if this can be made effective a halt on the depletion of the ozone layer may be achieved. Unfortunately there are other man-made chlorine compounds which may have a similar effect, but the whole question is one about which much remains to be understood.

Global Warming

For many millions of years our Sun, like other suns, has been getting hotter. But don't get alarmed. We are talking of thousands of millions of years, so the phenomenon is not likely to be significant for us at the moment.

The heat from the Sun is produced by the fusion of hydrogen and helium atoms which you will remember produces prodigious amounts of energy (Chapter IV). When the hydrogen and helium atoms are used up, other atoms such as carbon start to fuse and the Sun becomes hotter still and eventually expands and we, and all the neighbouring planets will be consumed. As I have said, hopefully, not for many millions of years.

If this is so, how is it that the surface of the Earth has not got appreciably hotter during this time? That this is so has been confirmed by the investigation of fossil forms produced millions of years ago. Although there have been various "Ice Ages", these could have been produced by quite a small variation in the temperature of the Earth's surface. So why has the Earth not got hotter?

The answer to this question is in the atmosphere - that gaseous envelope which surrounds and protects us.

The Moon, which has no atmosphere has as a result a surface temperature of about 100 deg. C (the boiling point of water) during the day which falls to minus 150 deg. C at night.

Our atmosphere, as I have mentioned before, consists of oxygen and nitrogen and a small proportion of carbon dioxide and other gases. Carbon dioxide is a colourless gas which you cannot see because visible light passes through it without being absorbed. However this is not the case for radiation of other wavelengths.

Nearly half the radiation we receive from the sun is in the visible region - 400 to 700 nm (See Table 1 in Chapter I) and strikes the Earth's surface, warming it. This heat is then radiated back to space but at a much longer wavelength between 1200 and 1800 nm and is absorbed by the carbon dioxide which becomes hotter. Some of this heat is transferred back to the surface which gets still hotter and the process repeats itself. The carbon dioxide is acting as a heat blanket.

The heat on the surface causes water to evaporate, forming water vapour which also absorbs radiation but at still longer wavelengths. This also traps the heat.

This then is the "Greenhouse Effect" - so called by the analogy with a greenhouse where the glass passes the shorter wavelengths but absorbs and traps the longer ones - so heating the greenhouse.

We are told that the original atmosphere of the Earth was nearly all (98%) carbon dioxide. If this had persisted we should have been like the plant Venus which today has an atmosphere consisting of 95% carbon dioxide and a surface temperature of nearly 500 deg. C - so if there was any lead on Venus it would have rivers of molten lead and not water!

Fortunately for us our atmosphere changed and there is only a trace of carbon dioxide left. Why was this? Some was absorbed in the oceans but the main route was a devious one through rocks which contained calcium and silicon and minute marine organisms. The calcium combined with the carbon dioxide to form limestone or chalk which we see today. These organisms as well as all plant life take in carbon dioxide from the air but when they decompose the carbon combines with oxygen to form carbon dioxide again so there is no overall reduction.

Sometimes, over millions of years, these organisms and plants are buried and cannot obtain any oxygen. They also become heated by the heat produced in the centre of the Earth by the radioactive decay of atoms which is going on there all the time. The results, after many more millions of years, is to form petroleum, coal, or natural gas - all compounds of carbon and hydrogen. When burnt they all produce carbon dioxide which then enters the atmosphere.

Man has been releasing carbon dioxide into the atmosphere for centuries, ever since he discovered fire, but only in the last century and as a result of the Industrial Revolution has the carbon dioxide been produced at ever increasing rates, mainly from power stations burning "fossil fuels" and from the internal combustion engine which powers our motor cars and aircraft.

What will happen to this carbon dioxide is a matter of some conjecture. Some will be absorbed in the oceans which are effectively huge sinks. Those marine organisms may increase their capacity to mop it up. Green plants also absorb carbon dioxide and as a result increase in size as you will know if you have a greenhouse and have an oil burning stove in it which is emitting carbon dioxide.

The cutting down of the forests which happens in South America, for example, will cause more carbon dioxide to be released when the wood is burnt or decays and with fewer trees there will be less of their foliage to absorb the carbon dioxide.

In spite of the difficulties of accurate measurements over a global area, we are now certain that the amount of carbon dioxide in the atmosphere is increasing. Not only that but other chemicals have a similar effect to carbon dioxide.

The CFCs, we have just discussed in connection with the ozone hole are a thousand times more effective as a heat blanket than carbon dioxide, although present in much smaller quantities.

So also are the oxides of nitrogen released from the soil as a result of the application of large quantities of nitrogenous fertilisers.

So also is the methane produced in the stomachs of cattle and similar animals (including humans) which we have mentioned before in discussing the Ozone Hole.

What concerns us is whether it will affect our weather and our climate significantly. This is difficult to predict as there are other influences. For example, volcanic eruptions, such as Krakatoa in the last century and Mount St Helens in this, which threw up vast quantities of dust into the atmosphere causing abnormal weather conditions as well as the magnificent sunsets we frequently see after them. It may be that the burning oil wells arising from the Gulf War may have a similar effect.

What will these effects be? Plants may grow more strongly as we have already indicated due to the increase in carbon dioxide.

However, an increase in the average temperature of the Earth's surface will be the more dramatic. There have been many predictions. It is considered at present that we may have an average rise of 2 deg. C by the middle of the 21st century and perhaps 5 deg. C by the year 2100. A rise of 1 deg. C could perhaps be contained but 4 deg. C could be catastrophic.

The Polar Ice Caps would melt exposing dark areas which would then warm more quickly from the incoming radiation than would the white ice and snow. This would accelerate the melting, producing water which would cause the level of the Oceans to rise to perhaps 165 ft. above present sea level, inundating low lying land and cities.

A rise of 2 deg. C would be bad enough. Not only would there be some rise in sea levels but the climate would change. Parts of the world would become warmer and parts colder, along with a change in the rainfall pattern, upsetting the conventional agricultural crops in different areas.

Well what can we do?

Very little about the carbon dioxide already released, but if we are to minimise its release in the future, drastic action will have to be taken.

To absorb the carbon dioxide from the fossil fuelled power stations could be counter-productive. The most economic way is to absorb it in "lime water" - calcium hydroxide - but in order to obtain this we have to heat limestone (calcium carbonate) which gives up an equivalent amount of carbon dioxide to the atmosphere, so there is no gain.

To use other chemicals such as sodium hydroxide would be much more costly but would also require a huge expenditure of power to manufacture them and therefore more fuel to be burnt in the power stations.

A rather drastic solution has been proposed recently. It is that the carbon dioxide should be stored in disused oil or gas wells. How this would be done - whether the gas would be liquefied (requiring the expenditure of more energy) and whether it would be environmentally acceptable - is not clear.

The only sure way would be to avoid using these fossil fuels and to supply our energy requirements from alternative sources such as nuclear, wind, wave, geothermal or other means.

Automobiles produce not only carbon dioxide but also oxides of nitrogen. The latter can be removed by special fittings known as catalytic converters using precious metals such as platinum or palladium. In addition motor car engines are notoriously inefficient in the burning of their fuel. Much goes out via the exhaust and when acted on by sunlight turns the oxygen of the air into ozone - another very potent "greenhouse" gas.

Anyone who has been to Los Angeles or Mexico City will have seen the thick "smog" which hangs over the cities caused almost entirely by the ubiquitous motor car. So we must either give up our cars or make them more efficient. As an alternative we could develop electrically driven automobiles and produce the electricity for them without burning fossil fuels

Perhaps an idealistic approach but if nothing is done the results may be disastrous.

Acid Rain

As well as contributing to global warming, some of the atmospheric pollutants have other effects.

The Sun warms the Earth and water evaporates - green plants also draw water from the soil by their roots and give it to the atmosphere from their leaves. The water vapour rises, then cools forming clouds from which rain falls to the Earth, completing what is called the Hydrological Cycle.

Water that has evaporated and then condensed should be pure, as is the "distilled" water for your car batteries (I forgot! your modern batteries don't need "topping up"!) but in passing through the atmosphere the water vapour collects various impurities, some natural and some man-made.

There are particles of soil, dirt, and soot. Some carbon dioxide is dissolved, but not enough to alleviate the Greenhouse Effect and the solution is slightly acid. The oxides of nitrogen are formed, as I have already mentioned, in motor car exhausts and from power stations and by the thousands of lightning flashes which occur all over the world. Although in the UK we may only have thunderstorms occasionally, they are going on all the time at various locations throughout the world and have significant effects. There are also volcanic eruptions which give out large quantities of sulphur in the form of sulphur dioxide.

These various "pollutants" react in different ways. The sulphur produces sulphuric acid and the nitrogen nitric acid. The carbon dioxide is itself a weak acid when dissolved in water. Not surprisingly these chemicals, all soluble in water, pollute the rain which falls on us.

The acidity or alkalinity of a solution is measured in terms of what is called its pH value (roughly from 1 to 10 - 1 being extremely acid and 10 extremely alkaline). Pure water is 7 and rain or snow an average of about 5 or 6 - so rain is itself slightly acid.

The rain falls, collects in lakes which feed rivers and so back to the sea. Freshwater lakes with a pH of 5 are fairly common. Much depends on whether they are in a peaty or a limestone area. If the latter then the calcium carbonate or bicarbonate present will neutralise the acidity giving a neutral pH of 7.

Even at a pH of 6 some species of fish will suffer and at 3.5 most fish will have been killed. Equally alkalinity can be deadly.

So, once again, why are we worried?

In the 1960s it was noticed that the stone surfaces of some buildings were becoming damaged and from the 1970s it was observed that the trees in the German and Scandinavian forests were suffering from an abnormal loss of foliage. It was realised that this could be caused by an increase in the acidity of the rain falling in that area. The origin of the acidity was immediately put at the door of sulphur dioxide emissions, predominately from fossil fuelled power stations.

Many coals and petroleums contain sulphur. When burnt either in power stations or domestically this appears in the atmosphere as sulphur dioxide, a colourless gas but which on reacting with water forms an acid solution (sulphuric acid).

It has been estimated that 40% of the sulphur in the atmosphere is man-made, the remaining 60% being from natural causes - volcanoes, and from marine organisms which produce methyl sulphide, etc.

In discussing the question of "Acid Rain" the media generally show pictures of large volumes of vapour arising from the cooling towers at power stations (we used to call them "Mae Wests" on account of their alleged similarity to the vital statistics of that one time well known actress) (Fig. 33).

You may have thought this was one of the pollutants causing acid rain and global warming. But it is not. The white cloud is water vapour or steam, just the same as the steam coming out of your boiling kettle.

Now you may well ask. Why is it necessary for all this steam to escape?

All power stations, whether they are coal, oil, or gas fuelled, or nuclear have to convert the heat which is produced from the burning of the fuel or from the nuclear reactor into electrical energy, which can be transmitted over long distances. In many cases this is converted back into heat in our homes.

The heat from the power station is first converted into high pressure and high temperature steam by means of a "boiler". This steam then passes through a turbine causing it to spin and drive an electrical generator - so the heat is now converted into electrical energy.

When the steam has passed through the turbine and done its work you could release it to the atmosphere but this would be very inefficient and would allow enormous clouds of water vapour into the atmosphere (far more than from the cooling towers). So it is more efficient to turn the steam back into water by cooling it. This is done by passing it through what is called a "condenser" which is cooled by cold water. This water is itself warmed in the process and would not be acceptable if it were returned to the environment as there is so much of it that it would warm up a river, for example, if it was passed into it. So this water has to be cooled and this is where the cooling towers come in.

The warm water is allowed to trickle down inside the tower while air travels up. This cools the water but some of the water does evaporate and produces the clouds of water vapour that you see.

The real pollution from the power stations, especially if they are fuelled by coal or oil (but not from nuclear stations which do not pollute) comes out of the tall chimney stacks. But you just can't see it as it is invisible. What you may see coming out of the chimney is a light plume which is probably fine dust. The clouds of steam are much more photogenic. You can't show the pollutants unless you have an infra-red camera.

The other main acid pollutant - oxides of nitrogen - forming nitric acid also arises from power stations due to the high temperatures of the furnaces which cause a reaction between the nitrogen and the oxygen of the air. The other source of nitric acid being from the exhausts from internal combustion engines.

To reduce the pollution is costly. Coal can be washed to reduce the amount of sulphur but involves a loss of ten times as much coal as sulphur removed. The preferred method is to pass the flue gases through a solution containing limestone (calcium carbonate) which is produced by quarrying in limestone areas. There are other methods but whichever is used they would all increase the cost of electricity and produce their own environmental problems.

There is still controversy as to the causes of "acid rain" and its effects and whether the proposed solutions, their costs and environmental side effects, are in the long run the best to be adopted.

I have tried to give a very "broad brush" view of the problems so that the reader can draw his own conclusions.

CHAPTER VIII

FUNDAMENTAL PARTICLES AND A BRIEF FORAY INTO COSMOLOGY

I have hesitated for some time as to whether I should write this chapter as it deals with a rather esoteric subject which may be difficult to understand but which I hope will be of interest.

So far I have tried to explain to someone with little knowledge of physics sufficient scientific background to enable him or her to understand what is written about the atom and the nucleus, and how it has led to the atomic bomb and nuclear power generation. I have also included an explanation of some of the developments arising from these matters in the fields of television, lasers, superconductivity and the environment.

Up to this point we have considered the proton and the neutron as the basic particles. Now we must realise that they are themselves made up of smaller particles. The speed of advancement in this branch of physics is so rapid that what I may now write may well be out of date by the time you read this.

I will first summarise the particles which are now considered to be "fundamental" and which make up the proton and the neutron. (We still consider the *photon* and the *electron* to be fundamental.)

The proton and the neutron are each composed of three particles called "*quarks*" - you will find some of the names in this chapter rather quaint I shall not try to explain how they were derived except that the modern physicist seems to have lost his love of words from the Classics.

Each quark has six "*flavours*" - up, down, strange, charm, bottom and top, and each flavour has three "*colors*" (I will use the American spelling as it has nothing to do with the colour we see with our eyes). A proton has two up quarks and one down quark whilst a neutron has two down and one up.

Another feature of these particles which make up our universe and, as we have seen in a previous chapter, can be considered either as a particle or a wave motion in space (arising from the quantum theory), is that they have a property which is called "*spin*". They spin round an axis - something like a spinning top - but it is how this axis is seen from an imaginary observer that decides their type. If it looks the same from any direction it is said to have spin 0. If it looks different from different directions it has spin 1. If it looks the same if you turn it through 360 degrees it has spin 2. The interesting thing is that some particles have to be turned through two complete revolutions to look the same. These have spin 1/2.

Now why have I gone to some length to explain this? Because all the particles in the universe can be divided into two groups. Those of spin 1/2 make up the whole of the matter in the universe whilst those of spin 0, 1, or 2 cause the forces which we experience.

To confuse you still further, these particles all have "*antiparticles*" with opposite properties such as electric charge. When they meet they mutually annihilate each other emitting a burst of energy.

There are various generic terms which you may come across. A "*fermion*", called after Enrico Fermi, covers all these particles. A "*hadron*" covers particles made up of quarks and antiquarks such as protons, antiprotons, neutrons, antineutrons, etc., whilst "*leptons*" include the others such as electrons, photons, etc., and their antiparticles.

Finally "*gauge bosons*" which are the particles associated with the transmission of forces, for example "*gluons*" and "*gravitrons*" which we shall meet in discussing the four fundamental forces. I hope that Table 3 (overleaf) will help you to put these things in perspective.

Now why should we bother with these particles with exotic names? Well, perhaps when you read about them in the Press or hear about them on television, you will have a better idea of what they mean. They do not produce any great technical "spin offs" in themselves although some of the results from their discovery may well lead to further developments in fusion power, electronics and the understanding of living matter. However they are essential if we are to strive, as scientists and philosophers have done over the centuries, to understand and solve the basic principles behind the world we live in and, more importantly, the universe of which we are such a minute part.

The study of "particle physics", as this science is called, and astronomy, cosmology, and the origins of the universe are all intimately connected.

Table 3

Table of Fundamental Particles

quarks		leptons	
up	down	electron	electro-neutrino
charm	strange	muon	muon-neutrino
top	bottom	tau	tau-neutrino

Gauge bosons				
gluon	photon	W	Z	gravitron

← increasing strength

Higgs boson
Higgs

(The "gluon" (the strong nuclear force) is 10^{38} times more powerful than the "gravitron" (gravity).)

The Fundamental Forces

The four forces have been considered briefly before: electromagnetic force, strong nuclear force, weak nuclear force and gravity.

We have met *electromagnetic* force in Chapter I where it caused the electromagnetic waves produced by the acceleration of an electric charge, producing both an electric and a magnetic field at right angles to each other and to the direction of propagation through space with the speed of light - 186,281 miles per *second*.

The electric charge can be on an electron, a quark or their antiparticles and can therefore be positive or negative. In large quantities of matter such as the Earth they cancel each other out and there is no nett electromagnetic force. At atomic dimensions this is not the case and we have seen that electrons are attracted by the nucleus and rotate round it. We have also seen when we discussed this that when an electron moved from one orbit to another, radiation was emitted as a burst of energy in the form of a photon.

Electromagnetic radiation covers the whole range of wavelengths, from radio, infra-red, visible light, ultraviolet, X-rays, gamma rays, and cosmic radiation. as you saw in Table 1 in Chapter I.

The *strong nuclear force* is the strongest of the forces, being a hundred times more powerful than the electromagnetic force. It only acts at extremely short distances such as are found within the nucleus and is the force which holds the quarks together in the protons and neutrons and is released when the nucleus is split or fissioned. It is considered that the binding energy of this force is produced by a particle called a "gluon" (see table 3, the name was produced presumably because it "glues" matter together!).

When particles are subjected to very high external energies which we shall discuss later in this chapter, the strong force becomes weaker and quarks and gluons behave like free particles.

The *weak nuclear force* also acts at extremely short distances and is a hundred thousand times weaker than the electromagnetic force and is responsible for the radioactivity produced by the radioactive decay of the elements.

Now we come to the *gravitational force* which is perhaps the most interesting of the four as we meet it in everyday life and also because it is perhaps not fully understood. It is the weakest by far of all the forces but acts over infinitely long distances as well as very short ones. It is now pictured as being carried by a particle called a "gravitron" which has no mass of its own and is called by the physicists a "virtual" particle which means to me that it only exists as a mathematical idea - although its effect is dramatic.

Newton's law of gravity stated that all bodies which had mass attracted each other with a force proportional to the mass of the bodies and inversely to the square of the distance between them. He also stated that gravity acted instantly throughout all space - i.e., it had an *infinite* velocity. This would mean it could be transmitted faster than the speed of light, which was contrary to Einstein's view that no information could be transmitted faster than this. We will return to Einstein's theories in the next section.

Space and Time

You will remember from Chapter I that early philosophers conceived that space was filled with an "aether". In the latter part of the nineteenth century attempts were made to measure the velocity of the Earth through this aether (if it existed) and what are known as the Michelson-Morley experiments were carried out. What Michelson and Morley reasoned was that if the Earth was moving through a "sea" of aether which was at rest, the velocity of light measured in one direction would be different from that measured at right angles (i.e. across the direction of motion - compare the velocity of a ship moving with a current with that when moving across a current). Their very precise measurements showed that there was no such difference and it had to be accepted that the aether did *not* exist.

(Don't get confused with the "Doppler" effect which is a change in *frequency* when the source of light or sound is moving with respect to the observer, e.g. the police car siren changes pitch when it is coming for you!)

We must now consider Einstein's theories of relativity, of which there are two - special and general - and to do so we must think of a *four* dimensional space as proposed by a Russian mathematician, Hermann Minkowski in 1907 - don't run away, I will try to make it as simple as possible!

We consider ourselves in a three dimensional space which means that we can describe an object by its length, breadth and height. The position of a ship at sea can be identified by its latitude and longitude. If it is an aircraft we must also know its height to say where it is. To me, three dimensions are fairly easy to visualise. But now we add a fourth - *time*. This is not so easy. It is true you can add the time of the aircraft at a certain place as another coordinate, and you can then calculate when you hope it will arrive. But this is not quite the point. The mathematicians just put it all into an equation but we mortals have to try to understand what it means. To us time flows forward. We know what happened five minutes ago but, although one might guess, we don't know what will happen in the next five minutes.

Minkowski postulated that all events in the universe occur in his four dimensional world that we now call "space-time". Einstein considered that this space-time was curved, not flat or rectangular as in our three dimensions. Light would travel in it by the shortest route which would not necessarily be what we call a straight line but rather like a "great circle" route on the surface of our globe. It was not until aircraft were able to make long distance flights that I realised they did not fly on a straight line from airport to airport but on a "great circle" route which was the shortest from one place to another.

The next thing he proposed was that this curvature of space-time occurred due to the presence of massive bodies such as the Earth or the planets. Two things follow from this, although perhaps they are not immediately obvious. First, that these masses will move as if attracted to one another and hence we have gravity. Secondly, light which travels by the shortest path in space-time will be deflected from its straight path (as it appears to us by an amount depending on the curvature of space-time as a result of the massive body.

There are as I have said two theories of relativity - general and special. I will deal with the General Theory first. This is supported by four "verifications" - three of which are from observations and the fourth partly from observations but mainly from theory.

The first point, that all bodies attract each other by gravity, was proved by observation of the orbit of the planet Mercury. This orbit was unusual in that instead of continuously retracing its path round the Sun, the orbit itself moves or precesses by a very minute amount (one revolution every three million years). According to Newton's laws, this orbit of the planet round the Sun should advance by 0.147 degrees every century but measurements had shown it to be 0.159 degrees - not a great difference you will say - but this difference of 0.012 degrees could only be explained by Einstein's theory. The fact that the orbits of the other planets do not precess significantly is because they are further from the Sun's gravity.

The second point, according to the theory, was that light passing by a massive body would be bent (attracted to it). We always think of light travelling in straight lines with no possibility of it being bent except by mirrors or refraction through the atmosphere. The only way to see if light is bent by passing close to a massive body such as the Sun is to measure the direction of a star or constellation from the Earth at night and then to measure the same stars when their light passes close to the Sun. This can only be done during a total eclipse of the Sun. On May 29, 1919, on the island of Principe in the Gulf of

Guinea, the astronomer Arthur Eddington measured the position of stars which had been located during darkness, as their light passed close to the Sun and showed the deflection to be as predicted by Einstein's theory.

The third point is to do with the vibration of atoms. We have seen, when we discussed spectra in Chapter II, that the colour of light emitted by a vibrating atom depended on the frequency of its vibration. Einstein predicted that an atom subjected to an intense gravitational field would vibrate slower than normal. So an atom in the Sun for example would be expected to vibrate slower than a similar one on Earth. We can prove this fairly simply. Taking the sodium atom as an example, it emits a yellow light of a certain frequency. There are many sodium and other atoms in the Sun. When we measure the frequency of the light from a sodium atom in the Sun we find it is slightly less than on Earth. It has moved towards the red end of the spectrum (lower frequency - therefore longer wavelength). This is know as "Gravitational Red Shift". The "Red Shift" that you read about in astronomy is caused by the light emitted by a moving object - a star - appearing more red when it is receding from the observer than it would if it were stationary or approaching (the Doppler effect). The fact that light received from stars in distant galaxies do suffer a "red shift" is taken as evidence that the universe is expanding.

The last point is perhaps more theoretical and is concerned with "black holes". In 1958 David Finkelstein had considered a theory based on Einstein's general theory that there could exist what he called a "one way membrane" where under extremely dense gravitational fields nothing could pass through it and even light could never escape. In a black hole all the laws of physics break down and even space and time disappear. Anything sucked into it can never escape except that it is suggested that it could be ejected from its other side into "another universe" - such is the subject of space fiction! To learn more about "black holes" I recommend to the reader Roger Penrose's book "The Emperor's New Mind".

Now to consider the Special Theory. Very briefly this is to do with the observation of phenomena by observers who are at rest and by those who are in motion and secondly that the velocity of light is the same for all observers irrespective of their own velocity and is a constant which cannot be exceeded.

I suppose that when we think of Relativity, one of the most popular images is to think of ourselves in a train at Euston station with another train on an adjoining platform. When this train starts and we look at it out of our window we don't know whether it is ourselves or the other train which is moving - this is assuming that British Rail can still start their InterCity trains without a jerk! What it really means is that speed - which is a measurement of both length and time - is relative to and depends on the speed or motion of the person observing them. There is no *absolute* time which is the same for all observers.

If you imagine that the observer was making his observations not from the train but from the back of a neutron, for example, (which of course is not possible in practical terms) and if this neutron was moving with a speed approaching that of light, then you would expect the velocity of light that you measured would vary with the speed of the neutron. So that if you were moving away from the light source, the speed of light would appear to be greater than 186,281 miles per second. Einstein says that this is just not possible. The only other explanation is that the neutron itself has changed - it has contracted in the direction of its motion. It has got smaller and therefore it has lost some of its mass. So Einstein made his second great prediction that mass and energy were related - if you lost mass you gained energy and vice versa and so we get the famous relation - Energy equals Mass multiplied by the velocity of light squared ($E = MC^2$) - if it were not true, there would have been no atomic bomb nor nuclear power.

Cosmic Rays

It is difficult to decide when cosmic radiation was first detected. In the 19th century experimenters found that their electroscopes (instruments used for detecting electric charge) discharged on standing. They put this down to faulty insulation or to some residual radiation in the material in the vicinity. In 1909, Professor Rutherford showed that a very penetrating radiation existed which discharged electroscopes even when they were shielded with large masses of lead. In 1932, Professor August Piccard took these instruments in a balloon to a height in the atmosphere which had never been reached by man before - 53,600 feet. He showed that the activity did not decrease with height as might have been expected if it was due to radiation from terrestrial sources. It must, therefore, have come from outside the atmosphere and was called cosmic radiation.

It was realised that shielding the instruments with lead was not very satisfactory as lead itself had a certain amount of residual radioactivity. In 1928, experiments were carried out by sinking the instruments to various depths in Lake Muir in California. This was at an altitude of 11,800 feet and the lake was fed with water from the snows which, as a result, had less residual radiation than the water from springs and streams. The results showed conclusively that the radiation being detected was reduced according to the depth of the instruments in the lake. At 73 feet the water was equivalent to the shielding effect of 6 feet of lead. At that time the hardest X rays were stopped by a half inch of lead and the hardest gamma rays known, by 12 inches. The rays, therefore, were more penetrating and of shorter wavelength than anything known at the time.

The origin of these rays was a matter of conjecture. In the 1930s, Professor Lindemann and Dr Dobson of Oxford University had a theory that they might be caused by thunderstorms. The electrical potential in these storms was sufficient to account for the high energy of the cosmic rays. In support of this theory there was a known variation in the frequency of thunderstorms throughout the world and the variation in cosmic ray intensity could also be recorded. Although a very slight correlation was found it was accepted that the radiation came from outside the Earth's atmosphere.

So what are cosmic rays? We now consider that 90% are produced by high energy protons originating from the explosion of stars in the distant parts of the universe. On striking the upper layers of our atmosphere they react with our air molecules to produce high energy electrons and other particles which we detect at ground level.

An interesting verification of one of the aspects of Relativity Theory arises from these phenomena. One of the particles produced is the *muon* (see Table 3). Muons have a mass 200 times that of an electron and can be produced in the "particle accelerators" I shall describe shortly, but they have a very short life which would not be long enough for them to reach the Earth's surface from the upper atmosphere before they decayed. However, they *do* reach the surface as they can be detected there.

So what is the explanation?

The answer seems to be that the muon is travelling at 99% of the speed of light. At this speed, according to relativity theory, time is shortened from the point of view of an observer on the Earth's surface and the muon appears to live longer (in fact seven times longer!) enough for them to reach the surface, although from the muon's point of view they still decay after their very short life.

This must seem a piece of utter nonsense to our common sense minds, but it is some experimental evidence which supports the Special Theory of Relativity and could be significant in space travel.

Whilst talking of the muon, it has been said that seven of them are passing through our bodies every second! If you remember, I said in Chapter I that we were nothing but space and little bundles of energy, so there is plenty of room for muons and other particles to pass through us. As a result of all these particles reaching the Earth's surface very high energy, short wavelength, gamma radiation is produced, being the cosmic radiation we experience. Further investigation showed that the intensity of this radiation coming from the West was greater than that coming from the East. This indicated that the particles striking our atmosphere were being deflected by the Earth's magnetic field and the direction of the deflection confirmed that they were positively charged and were therefore likely to be protons.

In the 1930s and '40s it was realised that these exotic particles were being produced by enormous energies never before produced on Earth. This led to the construction of large and very powerful particle accelerators to try to produce them artificially and to investigate their properties.

There was another development in the 1950s when the Jodrell Bank Radio Telescope was built, (almost entirely through the initiative of the astronomer Sir Bernard Lovell). With this marvellous instrument it was possible for the first time to detect and measure the electromagnetic radiation caused by the explosion of stars many millions of light years away and producing the showers of protons which are making the cosmic radiation we are now detecting.

The question of precisely where these particles come from is a matter of some controversy and is connected with the origins of the universe. That they come from outer space as well as our own galaxy is accepted. If the nearby region of the universe contains large quantities of anti-matter (you will remember I said that all the sub-atomic particles had antiparticles) we should expect large bursts of energy when they annihilated one another. But this does not appear to happen or to have happened in the distant past. So why is there no anti-matter present naturally in our universe today? It is suggested that equal amounts of matter and anti-matter were created in the "Big Bang" theory of the origin of the universe. For some reason, whilst most of the matter and anti-matter annihilated each other, a small amount of matter remained to form the universe as it exists today - or we should not be here!

Chaos

I cannot do more than touch very briefly on this subject which has become the latest tool in the scientist's workshop, to be considered as important as the theories of Newton, Einstein and Planck.

Chaos appears to me to be the science of unpredictability. It is not the unpredictability of the uncertainty principle which we have already met on the microscopic scale but the difficulty of predicting events on the larger or as we now say the macroscopic scale.

Science tries to explain cause and effect. The universe should run on a predetermined pattern according to these laws. To a great extent it does until suddenly, for no apparent reason, it breaks into chaos. For example, a smooth flowing stream suddenly becoming turbulent, the cigarette smoke from an ashtray, at first rising in a steady column, suddenly spreading into random tendrils. Until recently it was considered to be too complicated and difficult to study the underlying reasons for chaotic behaviour although hydraulic engineers and aircraft designers knew well enough what were the results when it occurred.

The original experiment which started the chaos theory is attributed to a meteorologist - Edward Lorenz - in 1961 when he was running his elementary computer to produce an artificial forecast of future weather. The information that he fed in was in the form of numbers to six places of decimals. On a subsequent run, to save time, Lorenz rounded the numbers to two places (thus 3.01234 became 3.01 and 3.01678 became 3.02). When he ran the computer with these numbers and compared the result with the previous ones, he found that while the two runs agreed to start with they suddenly and unpredictably became chaotically different. What had happened was that in "rounding" the numbers he had introduced a very small change in the initial conditions. You might have expected such a small change at the beginning to produce a small change throughout the run but instead it produced at a certain time a large and chaotic one.

This is now popularly known as the "Butterfly Effect" - it is suggested that a small disturbance in the atmosphere on the other side of the world such as the flapping of a butterfly's wings in Amazonia can be responsible for violent storms in Europe.

Later Lorenz carried on experimenting with numbers, apart from meteorology, and found a remarkable fact. If he had three variables and plotted the results on paper, he obtained an intricate pattern with two "eyes" and lines which never crossed but which unpredictably switched from one "eye" to the other. What he had found, in the new terminology of chaos, was a "strange attractor" or "fractal". (See Fig. 34.)

The fractal that Lorenz had produced was a curve enclosing a finite area but of infinite length. The example that is generally given is the measurement of the length of the coastline of an island. If you walk round the island with strides of, say, one metre, you will get a certain length, but suppose you walk with smaller steps and then into every inlet you will get a longer length. If you then go round every stone and eventually every molecule or atom you will get an infinite length. In fact, you can define a fractal as a way of seeing infinity!

Fractals can be produced mathematically and wonderfully beautiful patterns appear which replicate themselves to infinite smallness. You may have seen pictures of the Mandelbrot set which are generated from a comparatively simple mathematical equation.

Chaos theory is essentially mathematical but it can be applied to the real world to produce some amazing results in the fields of population growth, weather forecasting, biology and many more. To know more about it I would recommend the reader to James Gleick's book "Chaos".

Particle Accelerators

In Chapter III I described the "racetracks" which were used to separate the uranium isotopes. They had been developed by Ernest Lawrence from his first circular particle accelerator at the University of California in 1930 which had a diameter of 13 cm and accelerated protons to an energy of 80,000 electron volts.

You will hear a lot about electron volts. It is a measure of *energy* and is the energy acquired by an electron or any particle whose voltage is changed by one volt. Not surprisingly, physicists also use it as the measure of the *mass* of a particle - mass and energy, as you now know, being interchangeable ($E=MC^2$).

The energies we shall be talking about are very large - millions of electron volts (Mev) but they apply to individual particles and require enormous machines to produce them. In a chemical reaction such as the burning of carbon - one atom of carbon burning to carbon dioxide would only produce about five electron volts of energy, but remember it is only the one atom not a ton of coal. If that one atom were to disappear completely, it would produce about 3×10^{10} E/v of energy. Multiply this by the number of atoms in a ton of coal and you see what a prodigious amount of energy is available!

Since then many particle accelerators have been made with ever increasing energies to investigate the sub-atomic particles and to discover what they are made of. In California the Stanford Linear Accelerator Centre (SLAC) which was 2 miles in circumference, discovered the existence of quarks and other sub-atomic particles.

In Geneva, an organisation known as CERN (Centre Européen Recherche Nucléaire) built an accelerator nearly three miles in length (remember Lawrence's was only a little over a metre), and working with high energy collisions of protons and anti-protons discovered the W and Z particles (see Table 3).

The original equipment, known as the Super Proton Synchroton (SPS) was housed in a tunnel cut out of the rock under the borders of France and Switzerland to protect the equipment from outside forces and vibration. In excavating the tunnels to house the SPS, the engineers used instruments so sensitive that they could hear the vibrations from the waves of the Atlantic Ocean breaking on the coast of the Bay of Biscay.

Later a new tunnel was built adjacent to the original but nearly seventeen miles in circumference. This was known as the Large Electron Positron Collider (LEP) and accelerated particles to enormous speeds and smashed them into other particles and observed the results.

The latest equipment, known as the Large Hadron Collider (LHC) is housed in the same tunnel (it had been made large enough during construction in case this development was required) and will hurl electrons and positrons (the anti-particle of the electron) round it at ever increasing speeds in two beams in opposite directions kept apart by magnetic forces. When after many passes round the seventeen mile track the particles will have reached a speed very close to that of light and with an energy of several million million electron volts, they will be brought together and collide head on. Their energy will be converted into matter much as it is thought happened at the commencement of the universe in the "Big Bang". The investigation of the debris from these encounters should give us further insight into the forces which control the universe, and in particular the search for the Higgs Boson which theoretically should exist.

Of course all this costs money - very big money! - and some people will question if it is worth it. But man's insatiable quest for discovery will probably see that it is found.

(See Fig. 35.) Whilst CERN is a European project, the Americans are also "getting in on the act" with a huge superconducting supercollider (SSC) 54 miles in circumference, in Texas. It is reputed to be able to collide particles at 20 million million electron volts. But it has not yet been built.

At the energies envisaged with these machines, the fundamental forces we have been describing tend to become unified into one consistent mathematical description, known as the Grand Unified Theory or GUT. If it is achieved it should, according to the scientists, be able to explain everything.

Let us hope it does not release another genie from the cosmos.

CHAPTER IX

EPILOGUE: WHERE DO WE GO FROM HERE?

Everything we have so far discussed in physics, and physics can be described as the science of the universe, is based on the laws and theories of Newton, Einstein, quantum mechanics and chaos which are the ones I have tried to explain. But how do we know them to be true? And what is "true"? We have tested them against our physical observations with which they agree and so we assume them to be true. It has been said that mathematicians can see intuitively that a proposition is true although they cannot prove it. Perhaps as the poet has said "Truth is beauty, beauty truth. That is all you know and all you need to know".

In seeking this truth we may well ask the questions: how did the universe start? Is it the only one? And why are we here? The generally accepted answer to the first question is the "Big Bang" theory when in an inconceivably short period of time (one second divided by 10 with 100 zeros after it) energy was transformed into matter which evolved into the universe. But what caused this to happen? Not only did the Big Bang start the universe but it also started "time" but if we assume something caused the Big Bang we would expect there to have been time before it. The quantum theory tells us that we cannot predict events, only the probability of them happening. The Big Bang could have happened by a "probability" that energy at that time would be transformed into matter. How did that probability happen? Einstein has said "God does not play dice" but perhaps he decided in this case how the dice should fall. Where did that energy come from? Has it always existed? And we may then ask another question - what is "always"? Again, how did the laws of quantum mechanics arise and not some other laws? So many unanswered questions have resulted in the search for the Grand Unified Theory (GUT) - a theory to explain everything. If we do achieve a GUT what will it tell us? Why our universe came into being, and is it the only one possible - are we alone? Stephen Hawkins in his book "A Brief History of Time" suggests there is no beginning and no end but just "is" and there is something which he defines as "imaginary" time - a concept that only mathematicians can conceive. When we ask ourselves the question as to why we are here, why we exist at all on this universe there is a curious fact concerning the balance of matter and radiation within the universe. This has been shown to be about 10^9 to 1 in favour of radiation, measured in terms of the number of photons, to the number of neutrons and protons. For reasons which I will not go into it has been shown that if this ratio was very much different we should not be here. So what made this ratio so critical to our existence happen? The GUT if we find it cannot tell us why the laws arose in their particular form and no other. If we cannot answer these questions we may consider it as being due to "The Hand of God" and we must put this down to our own personal "God" of Love and Justice which we must have faith to believe in.

I like the following quotation from Sir Bernard Lovell's book "Astronomer by Chance":

"Finally we shall reach a stage where theories based on our present conception of physical laws have nothing further to say. At this point we pass from physics to metaphysics; from astronomy to theology, where the corporate views of science merge into the beliefs of the individual... We reach the great barrier of thought because we begin to struggle with the concepts of time and space before they existed in terms of our everyday experience. I feel I have driven into a great fog barrier where the familiar world has disappeared... The beginning itself is quite inaccessible."

So let us consider now what we ourselves are made of. A poet has said "We are such stuff as dreams are made of... and he may well be right. Physically very little. Millions of atoms which are practically all space themselves. Just little bundles of energy and when you break them down, as I have shown, they consist of quarks and now even smaller "*superstrings*". You might argue do we in the physical sense exist at all?

"What is man?... Stands he not thereby in the centre of immensities, in the conflux of eternities" - so said Thomas Carlyle in "Sartor Resartus". However, whatever we are, we are driven by the brain - a marvellous computer vastly more powerful and faster than any conceived so far and with memory cells many orders of magnitude greater than any produced up to the present. How is this computer programmed? That becomes a metaphysical problem and perhaps we shall never know.

Our brains are made up of neurons, which themselves are made up of atoms, which as I have pointed out above are nothing more than little points of electromagnetic energy, and if so is it possible that they radiate this energy. Radiation of all sorts of wavelengths - radio, cosmic, radioactive, are passing through our bodies all the time. Is it too fantastic to assume that the radiation from our brains is in some way modulated and transmitted to other brains! I personally have had no experience of thought transfer but I cannot see why it should not exist.

All this raises the great question - what happens when we die? Physically, the atoms and molecules of which we are made will be consumed and will appear as different molecules in the environment. It has been said that if you drink a cup of water it will contain at least one atom from the cup of hemlock which Socrates drank! I have not done the arithmetic, which makes the unlikely assumption that the atoms of Socrates hemlock are distributed uniformly throughout the atoms of our world. So I think we are pretty safe! If our brains do emit some form of radiation which is being propagated through space, which is the belief of some parapsychologists, why should this not continue after death? Is this our "Soul"? I don't think we shall ever have the answer to this or we should know the "Mind of God". We can only have faith if we are to believe that death is not the end.

THE SILICON CHIP

Sands of the desert, sands of the sea
Heat hot as hell
Pure as can be
Fine crystal growing
Then sliced like a loaf
Intricate pattern
Incredibly small
Doped with rare atoms
Emerges
The Chip.

What can it do?
Just a nought and a one
A there and not there
An and or not and, a neither nor.
It can add and subtract and everything follows
Remembers the answer.
It makes no decisions, accepts what it's told
It learns from its master
Cannot think for itself
Just a nought and a one
A there and not there

What is its use?
Turns on a tap, switches on switches off
Calculates this remembering that
Prints out the answer.
Takes over man's work and gives him more leisure
But what will he do with it?
That is the question
Just a nought and a one
A there and not there
It cannot do more than it's told
By Man and by God
Its Inventor.

APPENDIX

TABLE OF THE 103 KNOWN ELEMENTS IN
(A) ALPHABETICAL ORDER
(B) ORDER OF ASCENDING ATOMIC NUMBERS

(A) Alphabetical Order

Name	Atomic Number
Actinium	89
Aluminium	13
Americium	95
Antimony	51
Argon	18
Arsenic	33
Astatine	85
Barium	56
Berkelium	97
Beryllium	4
Bismuth	83
Boron	5
Bromine	35
Cadmium	48
Caesium	55
Calcium	20
Californium	98
Carbon	6
Cerium	58
Chlorine	17
Chromium	24
Cobalt	27
Copper	29
Curium	96
Dysprosium	66
Einsteinium	99

Erbium	68
Europium	63
Fermium	100
Fluorine	9
Francium	87
Gadolinium	64
Gallium	31
Germanium	32
Gold	79
Hafnium	72
Helium	2
Holmium	67
Hydrogen	1
Indium	49
Iodine	53
Iridium	77
Iron	26
Krypton	36
Lanthanum	57
Lawrencium	103
Lead	82
Lithium	3
Lutetium	71
Magnesium	12
Manganese	25
Mendelevium	101
Mercury	80
Molybdenum	42
Neodymium	60
Neon	10
Neptunium	93
Nickel	28
Niobium	41
Nitrogen	7
Nobelium	102
Osmium	76
Oxygen	8
Palladium	46

Phosphorus	15
Platinum	78
Plutonium	94
Polonium	84
Potassium	19
Praseodymium	59
Promethium	61
Protactinium	91
Radium	88
Radon	86
Rhenium	75
Rhodium	45
Rubidium	37
Ruthenium	44
Samarium	62
Scandium	21
Selenium	34
Silicon	14
Silver	47
Sodium	11
Strontium	38
Sulphur	16
Tantalum	73
Technetium	43
Tellurium	52
Terbium	65
Thallium	81
Thorium	90
Thulium	69
Tin	50
Titanium	22
Tungsten	74
Uranium	92
Vanadium	23
Xenon	54
Ytterbium	70
Yttrium	39
Zinc	30

Zirconium 40

(B) Order of Ascending Atomic Numbers

Those in bold appear in the text.
Atomic weights in last column.

Atomic Number	Name	Atomic Weight
1	**Hydrogen**	1.008
2	**Helium**	4.003
3	**Lithium**	6.94
4	**Beryllium**	9.01
5	**Boron**	10.82
6	**Carbon**	12.01
7	**Nitrogen**	14.008
8	**Oxygen**	16
9	**Fluorine**	19
10	**Neon**	20.18
11	**Sodium**	22.99
12	**Magnesium**	24.32
13	**Aluminium**	26.98
14	**Silicon**	28.09
15	**Phosphorus**	30.97
16	**Sulphur**	32.06
17	**Chlorine**	35.45
18	**Argon**	39.94
19	**Potassium**	39.1
20	**Calcium**	40.08
21	**Scandium**	44.96
22	**Titanium**	47.9
23	Vanadium	50.95
24	Chromium	52.01
25	Manganese	54.93
26	**Iron**	55.85
27	**Cobalt**	58.94
28	**Nickel**	58.71
29	**Copper**	63.55
30	Zinc	65.38
31	**Gallium**	69.72

32	**Germanium**	72.6
33	**Arsenic**	74.91
34	**Selenium**	78.96
35	**Bromine**	79.92
36	Krypton	83.8
37	Rubidium	85.48
38	Strontium	87.63
39	Yttrium	88.92
40	**Zirconium**	91.22
41	Niobium	92.91
42	Molybdenum	95.95
43	Technetium	(99)
44	Ruthenium	101.07
45	Rhodium	102.9
46	Palladium	106.4
47	**Silver**	107.9
48	Cadmium	112.41
49	Indium	114.82
50	Tin	118.7
51	Antimony	121.76
52	Tellurium	127.61
53	**Iodine**	126.91
54	**Xenon**	131.3
55	**Caesium**	132.91
56	**Barium**	137.34
57	Lanthanum	138.92
58	Cerium	140.13
59	Praseodymium	140.92
60	Neodymium	144.27
61	Promethium	(145)
62	Samarium	150.43
63	Europium	152
64	Gadolinium	157.25
65	Terbium	158.92
66	Dysprosium	162.46
67	Holmium	164.94
68	Erbium	167.2
69	Thulium	168.93

70	Ytterbium	173.04
71	Lutetium	174.99
72	Hafnium	178.49
73	Tantalum	180.95
74	**Tungsten**	183.85
75	Rhenium	186.20
76	Osmium	190.2
77	Iridium	193.1
78	**Platinum**	195.09
79	**Gold**	196.97
80	**Mercury**	200.59
81	Thallium	204.39
82	**Lead**	207.19
83	Bismuth	208.98
84	**Polonium**	209
85	Astatine	(210)
86	**Radon**	222
87	Francium	(223)
88	**Radium**	226.05
89	Actinium	227
90	**Thorium**	232.04
91	Protactinium	(231)
92	**Uranium**	238.03
93	**Neptunium**	(237)
94	**Plutonium**	242
95	Americium	(243)
96	Curium	(248)
97	Berkelium	(247)
98	Californium	(249)
99	Einsteinium	(254)
100	Fermium	(252)
101	Mendelevium	(256)
102	Nobelium	(253)
103	Lawrencium	(257)

BIBLIOGRAPHY

Allaby, Michael, *Green Facts. The Greenhouse Effect and other "isms"*
Appleyard, Brian, *Understanding the Present*
Casti, John, *Paradigms Lost*, Schribers
Collie, C. H., *Kinetic Theory and Entropy*
Davies, Paul, *The Mind of God*
Davies, Paul, *The New Physics*
Dixon, Bernard, *From Creation to Chaos*, Basil Blackwell
Edwards, Mark, *A Layman's Guide to the Cosmos*
Einstein, Albert, *Relativity*, Methuen
Feynman, Richard, *Genius and Modern Physics*
Gleick, James, *Chaos, Making a New Science*, Heinemann
Gribben, John, *The Hole in the Sky (Man's Threat to the Ozone Layer)*
Gribben, John, *In The Beginning*, Viking
Hall, Nina, *The New Scientist Guide to Chaos,* Penguin
Hawkins, Stephen, *From the Big Bang to Black Holes,* Bantam Press
Hawkins, Stephen, *Black Holes and Baby Universes,* Bantam Press
Lindemann F. A., *Quantum Theory*, Clarendon Press
Lovell. B, *Astronomer by Chance*
Midgley, Mary, *Science and Salvation*
Nelkon & Perkis, *Advanced Level Physics*
Penrose, Roger, *The Emperor's New Mind*
Proctor, R. F., *Nuclear Power - Its Development in the UK*, Gresham Press
Rhodes, Richard, *The Making of the Atom Bomb*, Simon & Schuster
Richards, H. J., *Death and After - What Will Really Happen?*
Russell, Bertrand, *ABC of Relativity*
Sakharov, Andrei, *Memoirs*, Hutchinson
Sullivan, J. W. N., *The Basis of Modern Physics*, Penguin
Sumner, David, *Radiation Risks*, Tarragon Press
Vaughan, Janet, *The Effect of Radiation on the Skeleton*
White, Michael, *A Life in Science*
Wiley, John, (B.M.A.) *Living with Risk*
Wilkinson, Denys, *The Organisation of the Universe*, Wolfson College Lectures, 1980

Wilson, F. A., *From Atoms to Amperes*, Babani
Zukav, Gary, *The Dancing Wu Li Masters*, Fontana

DIAGRAMS

TABLES

Picture Acknowledgements

Cover photographs	*Pamla Toler*
No. 14	*Apuan Alps- ENIT*
No. 23	*Tokamak - Stephen Williams*
No. 34	*Fractals and Mandelbrot Set*
No. 35	*CERN*

INDEX